A Computer Scientist's Guide to Cell Biology

T0186076

A Computer Scientist's Guide to Cell Biology

A Travelogue from a Stranger in a Strange Land

William W. Cohen
Machine Learning Department
Carnegie Mellon University

 Springer

William W. Cohen
Machine Learning Department
Carnegie Mellon University
Pittsburgh, PA 15213
USA
wcohen@cs.cmu.edu

Library of Congress Control Number: 2007921580

ISBN 978-0-387-48275-0   ISBN 978-0-387-48278-1 (eBook)

Printed on acid-free paper.

9 8 7 6 5 4 3 2 1

springer.com

To Susan, Charlie, and Joshua.

Table of Contents

List of Figures

Please visit the book's homepage at www.springer.com for color images of some figures.

Introduction

For the past few months, I have been spending most of my time learning about biology. This is a major departure for me, as for the previous 25 years, I've spent most of my time learning about programming, computer science, text processing, artificial intelligence, and machine learning. Surprisingly, many of my long-time colleagues are doing something similar (albeit usually less intensively than I am). This document is written mainly for them—the many folks that are coming into biology from the perspective of computer science, especially from the areas of information retrieval and/or machine learning—and secondarily for me, so that I can organize and retain more of what I've learned.

I find it helpful to think of "biology" in three parts. One part of biology is **information about biological systems** (for instance, how yeast cells metabolize sugar). This is the focus of most introductory biological textbooks and overviews, and is the essence of what biologists actually study—what biologists are trying to determine from their experiments. However, it is not always what biologists spend most of their time talking about. If you pick up a typical biology paper, the *conclusions* are typically quite compact: often all the new information about biological systems in a paper appears in the title, and almost always it can be squeezed into the abstract. The bulk of the paper is about **experimental methods** and how they were used—this, I consider to be the second part of "biology." The third part of "biology" is the **language and nomenclature** used, which is rich, detailed, and highly impenetrable to mere laymen. To read and understand current literature in biology, it is necessary to have some background each of these three parts: core biology, experimental procedures, and the vocabulary.

I like to think of the last few months as something like a field trip to a new and exotic land. The inhabitants speak a strange and often incomprehensible language (the nomenclature of biology) and have equally strange and new customs and practices (the experimental methods used to explore biology). To further confuse things, the land is filled with many tribes, each with its own dialect, leaders, and scientific meetings. But all the tribes share a single religion, with a single dogma—and all their customs, terms and rituals are organized around this religion. The highest goal of their religion is discover truth about living things—as much truth as possible, in as much detail as possible. This truth is "core" biology—information about living things. Knowing this "truth" is important, of course, but merely knowing the "truth" is not enough to understand a community of biologists, just as reading the Torah is not enough to understand a community of Jews.

In this document, I will provide a short introduction to "core" cell biology, mainly to introduce the most common terms and ideas. In doing so, I will occasionally oversimplify. This is deliberate. Computer scientists are used to analyzing complex systems by analyzing successively more complex abstractions, many of which are "real" (to the extent that anything computational is "real"): for instance, a push-down automaton is a generalization of a finite state machine, and both are useful for many real-world problems. One would like to operate in the same way in understanding biology, for instance, by first analyzing "finite-state" organisms, and then progressing to more complex ones. In biology, however, it is hardly ever the case that a clean and comprehensible abstract model perfectly models a real-life organism, so (almost) every simple general statement about how organisms function needs to be qualified—a tedious process in a document of this sort. I will also, by necessity, omit many interesting details, again deliberately. For a more comprehensive background on biology, there are many excellent textbooks, written by people far more qualified, some of which are mentioned in the final section of this paper.

After discussing "core" cell biology, I will then move on to discuss the most widely-used experimental procedures in biology. I will focus on what I perceive to be the high-level principles behind experimental procedures and mechanisms, and relate them to concepts well-understood in computer science whenever possible. Comments on nomenclature and background points will be made in side boxes.

How Cells Work

Prokaryotes: the simplest living things

One of the most fundamental distinctions between organisms is between the **prokaryotes** and the **eukaryotes**. **Eukaryotes** include all vertebrates (like humans) as well as many single-celled organisms, like yeast. The simpler **prokaryotes** are a distinct class of organisms, including various types of **bacteria** and **cyanobacteria** (blue-green algae). The best-studied prokaryote is *Escherichia coli*, or *E. coli* to its friends, a bacterium normally found in the human intestine. Like more complex organisms, the life processes of *E. coli* are governed by the "**central dogma**" of biology:

> "Bacteria" can refer to all prokaryotes, but more commonly refers to **eubacteria**, a subclass.
>
> DNA molecules are sequences of four different components, called **nucleotides**. Proteins are sequences of twenty different components called **amino acids**. Translation maps triplets of nucleotides called **codons** to single proteins: famously, nearly the same triplet-to-protein mapping is used by all living organisms.

DNA acts as the long-term information storage; **proteins** are constructed using DNA as a template; and to construct a particular protein, a corresponding section of DNA called a **gene** is **transcribed** to a molecule called a **messenger RNA** and then **translated** into a protein by a giant molecular complex called a **ribosome**. After the protein is constructed, the gene is said to be **expressed**. To take a computer science analogy, DNA is a stored program, which is "executed" by transcription to RNA and expression as a protein. The "central dogma" is summarized in Figure 1.

This same process of DNA-to-mRNA-to-protein is carried out by all living things, with some variations. One variation, which occurs again in all organisms, is that some RNA molecules are used directly by the cell, rather than being used only indirectly, to make proteins. (For instance, key parts of ribosomes are made of **ribosomal RNA**,

> Messenger RNA, ribosomal RNA, and transfer RNA are abbreviated as **mRNA**, **rRNA**, and **tRNA**, respectively. Another type of RNA, **small nuclear RNA** (**snRNA**), plays a role in splicing. A **gene product** is a generic term for a molecule (RNA or protein) that is coded for by a gene.

and mRNA translation also involves special molecules called **transfer RNAs**).

A second variation is that in the more complex **eukaryotic** organisms, mRNA is processed, before translation, by **splicing** out certain sub-sequences called **introns**. Surprisingly, the process of DNA-to-RNA-to-proteins is similar across all living organisms, not only in outline, but also in many details: scores of the genes that code for essential steps of the "central dogma processes" are highly similar in every living organism.

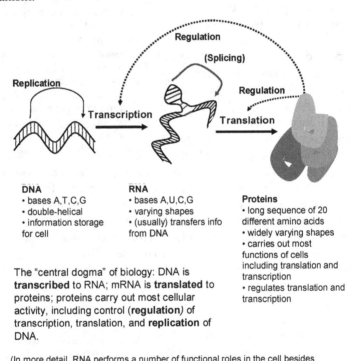

DNA
- bases A,T,C,G
- double-helical
- information storage for cell

RNA
- bases A,U,C,G
- varying shapes
- (usually) transfers info from DNA

Proteins
- long sequence of 20 different amino acids
- widely varying shapes
- carries out most functions of cells including translation and transcription
- regulates translation and transcription

The "central dogma" of biology: DNA is **transcribed** to RNA; mRNA is **translated** to proteins; proteins carry out most cellular activity, including control (**regulation***)* of transcription, translation, and **replication** of DNA.

(In more detail, RNA performs a number of functional roles in the cell besides acting as a "messenger" in mRNA.)

Figure 1. The "central dogma" of biology.

Prokaryotes are extremely diverse—they live in environments ranging from hot springs to ice-fields to deep-sea vents, and exploit energy sources ranging from light, to almost any organic material, to elemental sulphur. However, most pro-

> Membranes are composed of two back-to-back layers of fatty molecules called **lipids**, hence biological membranes are often called **bilipid membranes**.

karyotes are structurally quite simple: to a first approximation, they are simply bags of proteins. More specifically, a prokaryotic organism will consist of a single loop of DNA; an outer **plasma membrane** and (usually) a **cell wall**; and a complex mix of chemicals that the membrane encloses, many of which are **proteins**. Proteins are also embedded in the membranes of a cell.

A **protein** is a linear sequence of twenty different building blocks called **amino acids**. Different amino-acid sequences will fold up into different shapes, and can have very different chemical properties. Proteins are typically hundreds or thousands of amino acids in length. The individual amino acids in a protein are connected with **covalent** bonds,

> A **covalent bond** between two atoms means that the atoms share a pair of electrons. Weaker, inter-molecular forces include **ionic bonds** (between oppositely-charged atoms), and **hydrogen bonds** (in which a hydrogen atom is shared).

which hold them together very tightly. However, when two proteins interact, they generally interact via a number of weaker inter-molecular forces; the same is true when a protein interacts with a molecule of DNA.

One attractive force that is often important between proteins is the **van der Waals force**, a weak, short range electrostatic attraction between atoms. Although the attraction between individual atoms is weak, van der Waals forces can strongly attract large molecules that fit very tightly together. Another strong "attractive force" is **hydrophobicity**: two surfaces that are **hydrophobic**, or repelled by water, will tend to stick together in a watery solution, especially if they fit together tightly enough to exclude water molecules. Proteins, like the amino acids from which they are formed, vary greatly in the degree to which they are attracted to or repelled by water.

The importance of all this is that the interactions between proteins in a cell are often highly **specific**: a protein P

> A **bacteriophage**, or **phage**, is a virus that infects bacteria.

may interact with only a small number of other proteins—proteins to which some part of P "fits tightly." The chemistry of a cell is largely driven by these sorts of **protein-protein interactions**. Proteins also may interact strongly with certain very specific patterns of DNA (for instance, a protein might bind only to DNA containing the sequence "TATA") or with certain chemicals: many of the proteins in the plasma membrane of a bacteria, for instance, are **receptor proteins** that sense chemicals found in the environment.

Even simpler "living" things: viruses and plasmids

There are constructs simpler than prokaryotes that are lifelike, but not considered alive. **Viruses** contain information in nucleotides (DNA or RNA), but do not have the complete machinery needed to replicate themselves. Instead, they infect some other organism, and use its machinery to reproduce—just as an email virus uses existing programs on an infected machine to propagate. One well-studied virus is the **lambda phage**, which consists of a protein **coat** that encloses some DNA. The protein coat has the property that when it encounters the outer membrane of a cell, it will attach to the membrane, and insert the DNA into the cell. This DNA molecule has ends that attract each other, so it will soon form a loop—a loop similar to, but smaller than, the double-stranded loop of DNA that contains the genes in the host cell.

Even though this DNA loop is not in the expected place for DNA—that is, it is not part of any **chromosome** of the cell—the machinery for transcription and translation that naturally exists inside the cell will recognize the viral DNA, and produce any proteins that are coded by it. The DNA from the lambda phage produces a protein called **lambda**

> Most of the DNA in a cell is contained in **chromosomes**. In prokaryotes, a chromosome is generally a single long loop of DNA. Eukaryotic chromosomes have a more complex structure, and typical eukaryotes have several chromosomes.

integrase, which has the effect of inserting the viral lambda DNA into the host's chromosomal DNA. The cell is now a carrier of the lambda virus, and all its descendents will inherit the new viral DNA as well as the original host DNA. Eventually, some external event will make the virus become active: using the host's translation and replication machinery, it will excise its DNA out of the host's, create the materials (DNA and coat proteins) for many new viruses, assemble them, and finally destroy the cell's plasma membranes, releasing new lambda phage viruses to the unsuspecting outside world.

If DNA is the source code for a cell, then a lambda phage produces a sort of self-modifying program: not only is the central-dogma machinery of the cell appropriated to make new viruses, but the DNA that defines the cell itself is

> The **genome** is the "main" component of the genetic material for an organism— e.g., the chromosomal DNA for a eukaryote, or the nuclear DNA for a bacterium.

changed. This sort of self-modifying code is actually quite common, especially in eukaryotes, and the basic unit of such a change is called a **transposon**. There are many types of transposons—sections of DNA that use lambda-phage-like methods to move or copy themselves around the **genome**—and a large fraction of the human DNA consists of mutated, broken copies of transposons.

Even simpler than a virus is a **plasmid**, which is simply a loop of double-stranded DNA, much like the DNA inserted by a virus. Biologists have determined that there is nothing special about viral DNA that encourages the

> **Promoters** are DNA sequences that bind to the machinery that initiates the transcription of a gene. Without a valid promoter, a gene will not be expressed.

cell to use it: in particular, the machinery for DNA replication that naturally exists inside the cell will recognize a plasmid and duplicate it as well, as long as it contains, somewhere on the loop, the correct "instructions" for the replication machinery: for instance, one specific sequence of nucleotides called the **origin of replication** indicates where replication will start. Furthermore, the plasmid's DNA will also be transcribed to RNA and expressed, as long as it contains the proper **promoters**. In short, the DNA "program" in a plasmid will be "executed" by a cell, and the plasmid will be copied and inherited by children of a cell—just like the normal host DNA.

Plasmids are found naturally—they are especially common in prokaryotes. Like viruses, plasmids also occasionally migrate from cell to cell, allowing genetic material to pass from one bacterium to another.

(This is one way in which resistance to antibiotics can be propagated from one species of bacteria to another, for instance). There are also other plasmid-like structures that replicate in cells, but do not migrate from cell to cell easily—for instance, some yeast cells contain a loop of RNA that apparently encodes just the proteins needed for it to replicate.

All complex living things are eukaryotes

The class of **eukaryotes** includes all multi-celled organisms, as well as many single-celled organisms, like amoebas, paramecia, and yeast. Every plant or animal that you have ever seen without a microscope is a eukaryote. Surprisingly, in spite of their diversity, eukaryotes are quite similar at the biochemical level—there are more biochemical similarities between different eukaryotes than between different pro-karyotes, for example.

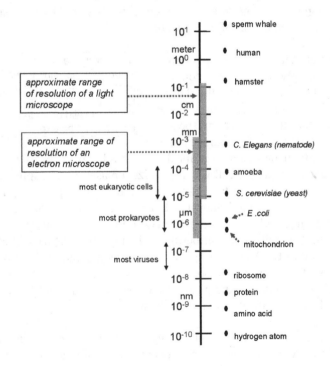

Figure 2. Relative sizes of various biological objects.

Eukaryotes are much larger and more complex than prokaryotes. The well-studied *E. coli*, for instance, is about 2 μm long, but a typical mammalian cell is 10–30 μm long, roughly 10–20 times the length of *E. coli*; this is about the same size ratio as an average-size man to a 60-foot sperm whale, or a hamster to a human. Figure 2 indicates the relative scale of some of the objects we have discussed so far.

Unlike prokaryotes, eukaryotes have a complex internal organization, with many smaller subcompartments called **organelles**. For instance, the DNA is held in an internal **nucleus**, specialized compartments called **mitochondria** generate energy, the **endoplasmic reticulum** synthesizes most proteins, and long protein complexes called **microtubules** and **microfilaments** give shape and structure to the cell. Figure 3 illustrates some of the main components of a eukaryotic animal cell.

Eukaryotes also use a more intricate scheme for storing their DNA "program." In prokaryotes, DNA is stored in what is essentially a single long loop. In eukaryotes, DNA is stored in complexes called **chromosomes**, wrapped around protein complexes called **nucleosomes**. The wrapping scheme that is used makes it possible to store DNA extremely compactly: for instance, if the DNA in a chromosome were about 1.5 cm long, the chromosome itself would be only about 2 μm long—four orders of magnitude shorter. Perhaps because of this ability to compact DNA, eukaryotes tend to have much larger genomes than prokaryotes.

In addition to containing much more DNA than prokaryotes, eukaryotes also postprocess mRNA by a process called **splicing**. In splicing, some subsections

> The parts of a gene that are "spliced out" are called **introns**. The parts that are retained are called **exons**.

of mRNA are removed before it is exported from the nucleus. Importantly, there can be multiple ways to splice the mRNA for a gene, so a single gene can produce many different proteins. This further increases the diversity of eukaryotes. Eukaryotes also have an additional set of mechanisms for regulating the expression of genes, because depending on its position relative to the nucleosomes, the DNA of a gene may or may not be accessible to the cell's transcription machinery.

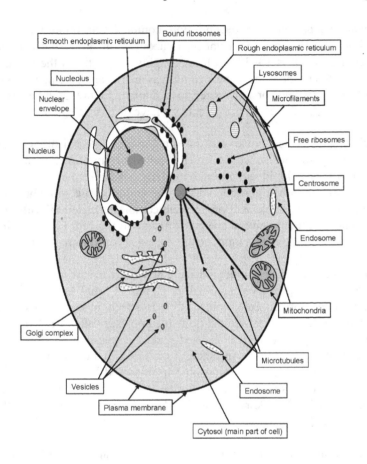

Figure 3. Internal organization of a eukaryotic animal cell.

It is believed that some of the organelles inside eukaryotes evolved from smaller, independent organisms that began living inside the early proto-eukaryotes in a symbiotic relationship. For instance, mitochondria might have once been free-living bacteria. One strong piece of evidence for this theory is that mitochondria (and also **chloroplasts**, an organelle found in plants) have their own vestigial DNA, which uses a different code for translating

> This theory of evolution is called **endosymbiosis**. A variety of modern endosymbionts exist, e.g., types of blue-green algae that live inside larger organisms. Some endosymbionts even contain a vestigial nucleus.

DNA triplets into amino acids than the scheme used by any modern organism.

Cells cooperate

Humans, elephants, mushrooms, trout and oak trees are all eukaryotes. Interestingly, at the molecular level, the cells in multi-celled eukaryotes are in many ways very similar to single-celled organisms. The various cells that make up a multi-celled organism will share the same DNA, but are **differentiated**, meaning that they express a different set of genes: for instance, a kidney cell will express a different set of genes than a muscle cell.

Cells in a multi-cellular organism also communicate, using a complex set of chemicals (mostly proteins) that are exchanged as **signals**, and received by **receptor sites** on the plasma membrane. Cells have many different ways of sending, receiving and propagating signals. The most common types of receptors are **ion channels**, which allow small charged particles to pass through a membrane, and **G-protein coupled receptors** (which are discussed more below).

Neurons make use of ion channels to send messages from cell to cell, and also to propagate messages along a cell. Neurons have many branch-like protrusions called **dendrites** that receive signals. Outgoing signals pass through another protrusion called an **axon**, which can be several feet in length. To send a signal down an axon, a chain of **voltage-gated ion channels** are used—channels that open in response to a voltage signal. Opening an ion channel means that ions rush into the cell (since the ions are normally in a higher concentration outside the cell than inside it), which causes another voltage spike—a spike strong enough to cause nearby ion channels to open...which causes those channels to generate voltage spikes, and stimulate their neighboring channels, and so on. The process is somewhat like a "wave" at a football game, as is illustrated in Figure 5.

Of course, in order for the neuron to be ready to transmit the next signal, it is also necessary that the channels close again after the "wave" has passed by. One scheme for handling this is shown in Figure 4: shortly after a channel opens, it closes, and immediately after closing, the channel is **inactive**—i.e., unable to respond to voltage signals. The inactive phase keeps the wave moving in a single direction, but also requires ion-channel protein complexes to have some sort of short-term memory. Thus, ion channels are not simple holes in a membrane—they are quite complex molecular machines. Their shapes are also highly optimized to allow only certain ions through—the most common ones for signaling between cells being sodium (Na) and potassium (K).

After responding to a voltage signal of this sort, a neuron has absorbed many sodium ions. These are rapidly removed by special molecular complexes that "pump" unwanted ions out. The high concentration of ions outside the neuron that is produced by the pumps provides the energy needed to propagate the voltage signal.

Another type of ion channel is opened by the presence of a chemical called a **transmitter** rather than by voltage. **Transmitter-gated ion channels** are used to send signals from one neuron to another, as is

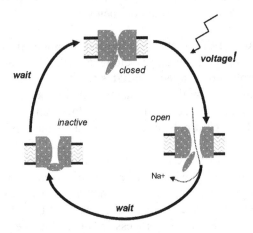

A voltage-gated ion channel with three states: **closed,** which opens in response to voltage; **open,** which allows ions to pass through; and **inactive,** which blocks ions, and does not respond to voltage. The open and inactive states are temporary.

Figure 4. Voltage-gated ion channels in neurons.

shown Figure 6. Transmitter-gated ion channels are also common parts of the membranes inside cells: for instance, there are many channels that release calcium (Ca) ions from inside the endoplasmic reticulum— where it is found in abundance—into the cytoplasm. As in the re-uptake

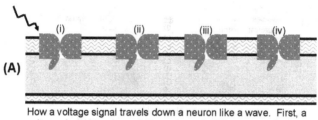

How a voltage signal travels down a neuron like a wave. First, a voltage signal hits channel (i), as shown in (A).

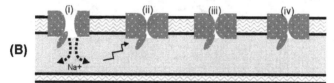

Then channel (i) opens, and ions rush in, causing a voltage spike that opens channel (ii), as shown in (B).

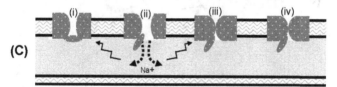

Then channel (ii) opens, sending voltage spikes to channels (i) and (iii), as shown in (C).

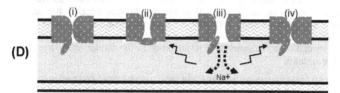

Next, channel (iii) opens, as shown in (D). Because (i) is inactive, it cannot open. Ion-produced voltage spikes are now sent to the inactive channel (ii) and the closed channel (iv). Channel (iv) will open next.

Figure 5. How signals propagate along a neuron.

process of Figure 6, calcium-based signals require a means of removing "old" signaling material; hence, calcium-based signaling is often associated with the protein **calmodulin**, which binds readily to calcium.

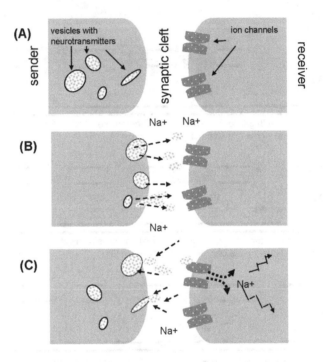

An example of a **transmitter-gated ion channel**. (A) shows the initial state. A substance used for signaling (for neurons, this is called a **neurotransmitter**) is held in vesicles by the sender cell. (B) In response to some internal change, the neurotransmitter is released. (C) Some of the neurotransmitter binds to ion channels on the receiver cell, and causes the channels to open. Most of the remainder of the neurotransmitter is re-absorbed by the sender cell, in a process called **re-uptake**.

A common neurotransmitter is **serotonin** (which is chemically related to the amino acid tryptophan). Many widely-used **antidepressants** (Prozac, Zoloft, and others) inhibit the reuptake step for serotonin, and are thus called **selective serotonin re-uptake inhibitors (SSRIs).** They cause serotonin to accumulate in the synaptic cleft, making it more likely that signals will propagate from cell to cell

Figure 6. A transmitter-gated ion channel.

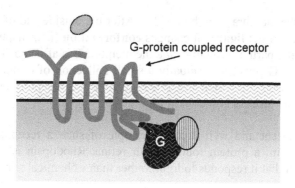

(A) A G-protein complex is bound to the G-protein coupled receptor on the inside of the cell. (There are many different types of G-proteins, and many types of receptors.)

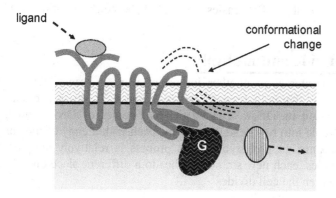

(B) When the receptor binds to the **ligand** molecule, then the entire receptor changes shape. As a consequence, the G-protein complex is altered: part of it is released, to propagate the signal elsewhere in the cell.

Figure 7. A G-protein coupled receptor protein.

Unlike ion channels, **G-protein coupled receptor proteins (GPCRs)** do not actually pass substances through a membrane. Instead, these receptors extend

A **ligand** is a molecule that binds to specific place on another molecule. The shape of a protein is called its **conformation**.

through the membrane on both sides. After the outside end of a **GPCR** binds to its target **ligand**, it changes **conformation** (i.e., shape) in such a way that a partner protein *inside* the membrane is affected. Typically, the partner **G protein** is actually a small collection of proteins bound together, some of which are released after the receptor detects the ligand. This process is shown in Figure 7.

One important and well-studied example of such a receptor protein is **rhodopsin**, a protein found in our retina. Rhodopsin is somewhat atypical in that it responds to light, rather than a chemical stimulus.

Receptor proteins (and signaling pathways in general) are extremely important clinically, because they provide the easiest way for drugs to affect an organism. In general, cells make it difficult for outsiders to move chemicals across the plasma membrane; if you want to make them behave, it is often easiest to exploit the cell's "existing API" of signaling responses.

Cells divide and multiply

Cells also interact in another important way: by reproducing. The simplest way that cells reproduce is by division. In this process a cell will duplicate its DNA, separate the two copies of DNA, and then finally divide into two "daughter" cells, each with a copy of the parent cell's **genome**. In prokaryotes, this process is relatively simple: the DNA divides, each new strand attaches to a different place on the cell wall, and then the cell divides.

Perhaps because the genetic material is organized into chromosomes, each of which must be duplicated and divided

> Cell division in eukaryotes is called **mitosis**.

among the daughter cells, the process of division in eukaryotes is quite complex. Eukaryotic cells progress through a regular cycle of growth and division called the **cell cycle**, consisting of four phases: **S phase**, during which DNA is synthesized; **M phase**, during which the actual cell division (mitosis) occurs; and two gap phases, **G1** and **G2**, which fall between M&S and S&M respectively. The M phase consists of a number of subphases: **prophase**, **prometaphase**, **metaphase**, **anaphase**, **telophase**, and **cytokinesis**, during each of which specific changes take place. (For instance, in metaphase, pairs of duplicate chromosomes are moved to the center of the nucleus.)

The cell cycle is orchestrated by a set of proteins called **cyclins** and **cyclin dependent kinases (Cdks)**. The many actual movements that take place in mitosis are produced by "molecular motor" proteins that interact with the cell's microtubules.

> A **kinase** is a protein that modifies another protein by adding a phosphate group. This process is called **phosphorylation.**

Like many things, this whole process becomes even more complicated when sex is involved. Organisms that reproduce sexually have two types of cells: **diploid** cells, which contain two copies of each chromosome, and **haploid** cells, which contain only one copy. Haploid cells are produced by a different type of cell division (called **meiosis)** which is illustrated below in Figure 8.

Only a single pair of chromosomes is shown in Figure 8, which simplifies the drawing. Unfortunately, considering a single pair of chromosomes also *overly* simplifies the process in an important way. Consider a diploid cell with N chromosome pairs: for convenience, call these pairs $(m_1, f_1), \dots (m_N, f_N)$. Meiosis will produce four haploid cells, each of which contains either m_1 or f_1, either m_2 or f_2, and so on; thus there are 2^N possible haploid daughter cells. The huge number of possible ways in which chromosomes can be divvied up during meiosis is reason why eukaryotic species, like ourselves, can be genetically diverse.

In fact, the number of possible haploids is much larger than this, due to **genetic recombination**, a process in which segments of DNA are "swapped" between chromosomes. As shown in Figure 8D, this typically occurs when bivalents are formed. These swaps, or **crossover events**, happen on average 2–3 times on each pair of human chromosomes.

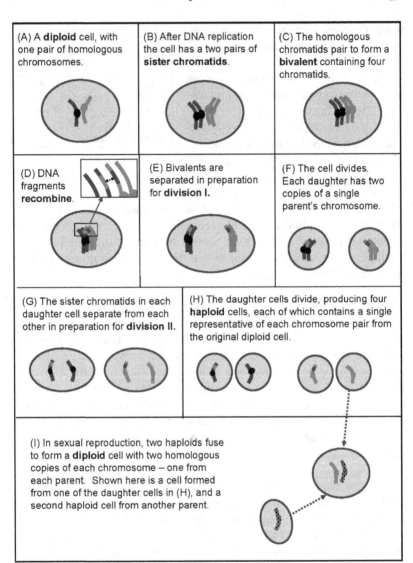

Figure 8. Meiosis produces haploid cells.

Diploid cells are more complex to study, if your goal is to understand which genes cause which effects, because the two copies of each gene need not be exact copies: instead, there can be slightly different DNA sequences that produce

> An organism with two copies of the same allele for a gene is **homozygous** for that gene. An organism with two different alleles for a gene is **heterozygous** for the gene.

similar gene products. The variant sequences are said to be different **alleles** of the gene. Often, only one of the alleles (the **dominant allele**) will be expressed, and the other **recessive allele** will be "hidden" (in the sense that its effects are masked).

In humans, there are only two types of haploid cells: egg cells and sperm cells. All other cells are diploid. A popular organism for genetic studies is **yeast**, a single-celled eukaryote that can grow and reproduce as a haploid, but can also reproduce sexually. There are no male or female yeast: instead the "sexes" for yeast are called **type a**, and **type α**. When yeast cells "want" to mate, they release a chemical called a **mating factor** (which, by the way, is detected by a type of G-protein coupled receptor). Yeast cells are not always receptive to mating signals—for instance, when there is plenty of food in the environment, they often "prefer" to eat. Sometimes, however, when a "Greek" type-α yeast cell detects a mating factor from a "Roman" type-a cell, it will start building a protuberance called a "schmoo tip"—a name derived from the classic "Lil Abner" cartoons by Al Capp. Eventually the "schmoo tips" of the parent cells grow together and the cells can fuse and mate, producing a diploid child.

Prokaryotes do not undergo meiosis, but they can exchange genetic material via plasmids. One special type of plasmid, called a **fertility plasmid** or **F-plasmid**, contains genes that enable an *E. coli* to initiate a process called **conjugation**. Bacteria containing the F-plasmid are called "male," and have the ability to construct a long tubular organelle called a sex pilus, which is used (you'll be relieved to read) as a sort of a grappling hook to grab another *E. coli* and bring it in close. The organisms then form a "conjugate bridge" and exchange genetic material—including the F-plasmid itself. Mating usually involves groups of 5–10 bacteria, and in the kinky world of the *E. coli*, all of them become "male" after conjugation, by virtue of their newly-received F-plasmid.

The Complexity of Living Things

Complexes and pathways

Although the basic mechanisms that underlie cellular biology are surprisingly few, there are many instances and many variations on these mechanisms, leading to an ocean of detail concerning (for instance) how the process of microtubule attachment to a centrosome differs across different species. Cellular-level systems, because they are so small, are also difficult to observe directly, which means that obtaining this detail experimentally is a long and arduous process, often involving tying together many pieces of indirect evidence. Most importantly, cellular biology is hard to understand because living things are extremely complex—in several different respects.

One source of complexity is the sheer number of objects that exist in a cell. At the molecular level of detail, there are thousands of different proteins in even the simplest one-celled organisms. These individual proteins can themselves be quite large, and assemblies of multiple proteins (appropriately called

> A **flagellum** is a whip-like appendage that certain bacteria have. It functions as a sort of propeller to help them move. An *E.coli* flagellum rotates at 100Hz, allowing the *E.coli* to cover 35 times its own diameter in a second.

protein complexes) can be extremely intricate. One notable example for bacteria is the "molecular motor" which spins the **flagellum**—an assembly of dozens of copies of some twenty distinct proteins that functions as a highly efficient rotary motor. (See Figure 9.) This motor is atypical in some ways—most protein complexes are less well-understood, and do not resemble familiar mechanical devices like turbines—but it is far from unrivaled in its size or in the number of protein components. (Ribosomes, for instance, are much larger.) Unraveling this type of complexity is part of the discipline of biochemistry.

A second type of complexity associated with living things are the complex ways in which proteins interact with each other, with the environment, and with the "central dogma" processes that lead to the production of other proteins. A *simplified* illustration of one of the best-studied such processes is shown in Figure 10, which illustrates how *E. coli* "turns on" the genes that are necessary to import lactose when its preferred nutrient, glucose, is not present. Briefly, the gene *lacZ* is regulated by two proteins (called *CAP* and the *lac repressor protein*), which function by binding to the DNA near the site of the *lacZ* gene, and a feedback loop involving lactose and glucose affect the relative quantities of CAP and the lac repressor protein; however, as the figure shows, the details of this feedback process are nontrivial.

Many cell processes involve this sort of "interaction complexity," and often the interactions are far from being completely deciphered, let

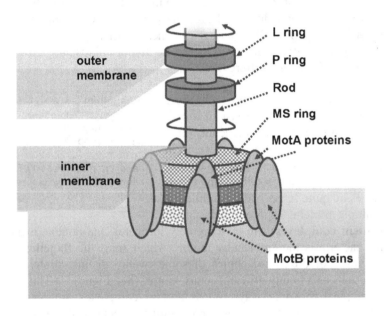

Structure of a bacterial flagellum (simplified). About 40 different proteins form this complex. The MS ring is made up of about 30 FliG subunits, and about 11 MotA/MotB protein pairs surround the MS ring. It is believed that these pairs, together with FliG, form an ion channel. As ions pass through the channel, conformational changes cause the MS ring to rotate, much like a waterwheel.

A similar "molecular motor" is used in ATP synthesis in a mitochondrion: rotation, driven by ions flowing through a channel, is the energy used to convert ADP to ATP. (See the section below, "Energy and Pathways").

Figure 9. The bacterial flagellum.

alone understood. Like the molecular motor that drives the flagellum, the chemical interactions in a cell have been optimized over billions of years of evolution, and like any highly-optimized process, they are extremely difficult to comprehend.

Individual interactions can be complicated

Networks of chemical interactions like the one shown in Figure 10 are also complex in a different respect: not only is there a complex network that defines the *qualitative* interactions that take place, the

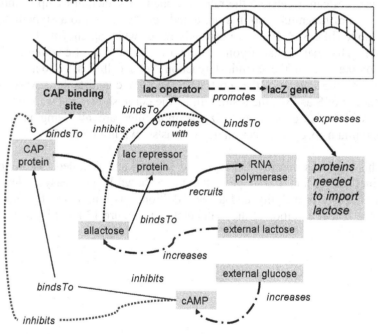

The *lacZ* gene is transcribed only when CAP binds to the CAP binding site, and when the *lac* repressor protein does not bind to the *lac* operator site.

This network presents simplified view of why *E.coli* produces lactose-importing proteins only when lactose is present, and glucose is not.

Figure 10. How *E. coli* responds to nutrients.

individual interactions can be *quantitatively* complex. To take an example, increases in glucose *might* increase the quantity of cAMP linearly—but often there will be complex non-linear relationships between the parts of a biological chemical pathway.

The reason for this is that most biological reactions are mediated by **enzymes**—proteins that encourage a chemical change, without participating in that change. Figure 11 gives a "cartoon" illustrating how an enzyme might encourage or **catalyze** a simple change, in which molecule *S* is modified to form a new molecule *P*. It is also common for enzymes to catalyze reactions in which two molecules *S* and *T* combine to form a new product.

Enzymes can accelerate the rate of a chemical reaction by up to three orders of magnitude, so it is not a bad approximation to assume that a change (like S → P above) can only occur when an enzyme E is present. This means that if you assume a fixed amount of enzyme E and plot the rate of the chemical reaction (let's call this "velocity," V) against the amount of the substrate S (and like chemists, let's write the amount of S as [S]), the result will be the curve shown below. Velocity V will increase until the enzyme molecules are all being used at maximum speed, and then flatten out, as shown in Figure 12.

This model is due to Michaelis and Menten and is called "saturation kinetics." In fact, the shape of the curve shown is quite easy to derive from basic probability and a few additional assumptions—the ambitious reader can look at the mathematics in Figure 13 and Figure 14 to see this.

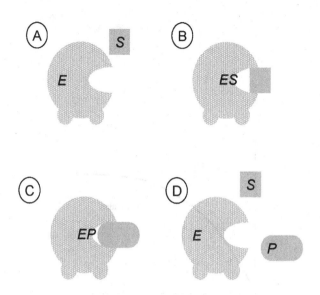

A cartoon showing how an enzyme catalyzes a change from *S* to *P*. (A) Initially, the enzyme *E* and "substrate" *S* are separate. (B) They then collide, and bind to form a "complex" *ES*. (C) While bound to *E*, forces on the substrate *S* cause it to change to form the "product" *P*. (D). The product is released, and the enzyme is ready to interact with another substrate molecule *S*. A chemist would summarize this as: E+S→ ES → EP → E+P

Figure 11. How enzymes work.

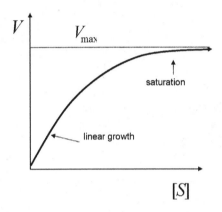

Reaction velocity with a fixed quantity of an enzyme
E, and varying amounts of substrate S. When little
substrate is present, an enzyme E to catalyze the
reaction is quickly found, so reaction velocity V grows
linearly in substrate quantity [S]. For large amounts
of substrate, availability of enzymes E becomes a
bottleneck.

Figure 12. Saturation kinetics for enzymes.

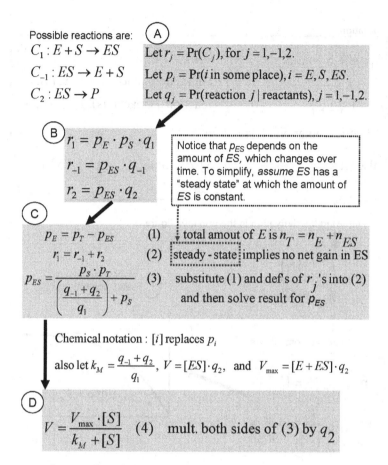

Possible reactions are:
(A)
$C_1 : E + S \rightarrow ES$ Let $r_j = \Pr(C_j)$, for $j = 1, -1, 2$.
$C_{-1} : ES \rightarrow E + S$ Let $p_i = \Pr(i \text{ in some place}), i = E, S, ES$.
$C_2 : ES \rightarrow P$ Let $q_j = \Pr(\text{reaction } j \mid \text{reactants}), j = 1, -1, 2$.

(B)
$r_1 = p_E \cdot p_S \cdot q_1$
$r_{-1} = p_{ES} \cdot q_{-1}$
$r_2 = p_{ES} \cdot q_2$

Notice that p_{ES} depends on the amount of *ES*, which changes over time. To simplify, *assume ES* has a "steady state" at which the amount of *ES* is constant.

(C)
$p_E = p_T - p_{ES}$ (1)   total amout of E is $n_T = n_E + n_{ES}$
$r_1 = r_{-1} + r_2$ (2)   steady - state implies no net gain in ES
$p_{ES} = \dfrac{p_S \cdot p_T}{\left(\dfrac{q_{-1} + q_2}{q_1}\right) + p_S}$ (3)   substitute (1) and def's of r_j's into (2) and then solve result for p_{ES}

Chemical notation : $[i]$ replaces p_i

also let $k_M = \dfrac{q_{-1} + q_2}{q_1}$, $V = [ES] \cdot q_2$, and $V_{\max} = [E + ES] \cdot q_2$

(D)
$V = \dfrac{V_{\max} \cdot [S]}{k_M + [S]}$ (4)   mult. both sides of (3) by q_2

See next figure for how to *interpret* Equation (4)....

Figure 13. Derivation of Michaelis-Menten saturation kinetics.

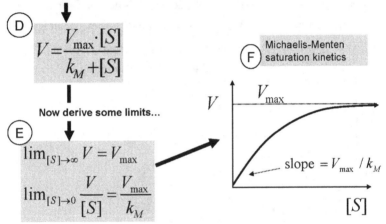

Notation:

$$C_1 : E + S \rightarrow ES \qquad r_j = \Pr(C_j), \text{ for } j = 1, -1, 2.$$

$$C_{-1} : ES \rightarrow E + S \qquad p_i = \Pr(i \text{ in random place}), i = E, S, ES.$$

$$C_2 : ES \rightarrow P \qquad q_j = \Pr(\text{reaction } j \mid \text{reactants}), j = 1, -1, 2.$$

Chemical notation : $[i]$ replaces p_i

also let $k_M = \dfrac{q_{-1} + q_2}{q_1}$, $V = [ES] \cdot q_2$, and $V_{\max} = [E + ES] \cdot q_2$

Following the derivation in the previous figure...

(D)
$$V = \frac{V_{\max} \cdot [S]}{k_M + [S]}$$

(F) Michaelis-Menten saturation kinetics

Now derive some limits...

(E)
$$\lim_{[S] \to \infty} V = V_{\max}$$

$$\lim_{[S] \to 0} \frac{V}{[S]} = \frac{V_{\max}}{k_M}$$

V | $V_{\max}$

slope $= V_{\max} / k_M$

$[S]$

The first limit shows that V, the velocity at which P is produced, will asymptote at V_{max}.

The second limit shows that for small concentrations of S, the velocity V will grow linearly with [S], at a rate of V_{max}/k_M.

Figure 14. Interpreting Michaelis-Menten saturation kinetics.

Enzymes with more complicated structures can lead to more complicated velocity-concentration curves, as shown in Figure 15. A typical example would be an enzyme with two parts, each of which has an **active site** (a location at which the substrate S can bind), and each of which has two possible **conformations** or shapes. One conformation is a fast-binding shape, which has a high maximum velocity $V_{maxFast}$, and the other is

> A molecule that is composed of two identical subunits is a **dimer**; three identical subunits compose a **trimer**; and *N* identical subunits compose a **polymer**. An enzyme in which binding sites do not behave independently is an **allosteric** enzyme; in the example here, the enzyme exhibits **cooperative binding**.

a slower-binding shape with maximum velocity $V_{maxSlow}$. The lower part of the figure shows a simple state diagram, in which: (a) both parts of the enzyme change conformation at the same time, (b) shifts from the slow to fast conformation happen more frequently when the enzyme is binding the substrate, and (c) shifts from fast to slow tend to happen when the enzyme is "empty," i.e., not binding any substrate molecule. In this case, as substrate concentration increases, the enzymes in a solution will gradually shift conformation from slow-binding to fast-binding states, and the actual velocity-concentration plot will gradually shift from one saturation curve to another, producing a **sigmoid** (i.e., S-shaped) curve—shown in the top of the figure. A sigmoid is a smooth approximation of a step-function, which means that enzymes can act to switch activities on quite quickly.

Sigmoid curves and network structures are also familiar in computer science, and especially in machine learning: they are commonly used to define **neural networks**. A neural network is simply a directed graph in which the "activation level" of each node is a sigmoid function of the sum of the activation levels of all its input (i.e., parent) nodes. It is well-known that neural networks are very expressive computationally: for instance, finite-depth neural networks can compute any continuous function, and also any Boolean function. Although I am not familiar with any formal results showing this, it seems quite likely that protein-protein interaction networks governed by enzymatic reactions are also computationally expressive—most likely Turing-complete, in the case of feedback loops. This is another source of complexity in the study of living things.

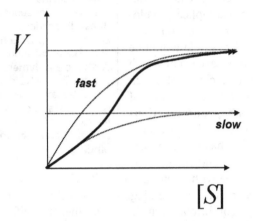

Allosteric enzymes switch from a slow-binding state to a fast-binding state, and tend to remain in the fast-binding state when the substrate *S* is common. Their kinetics follows a sigmoid curve.

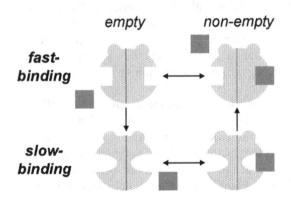

A typical allosteric enzyme: when one half is being used, the whole molecule tends to shift to the fast-binding state.

Figure 15. An enzyme with a sigmoidal concentration-velocity curve.

Energy and pathways

Enzymes are important in another way. Running the machinery of the cell requires energy. Most of this energy is stored by pushing certain molecules into a high-energy state. The most common of these "fuel" molecules is **adenosine**, which can be found in two forms in the

> More properly, ATP is combined with water to produce ADP plus inorganic phosphate, yielding energy: ATP+H₂0 → ADP + Pi. This reaction is called **hydrolysis**.

cell: **adenosine triphosphate (ATP)**, the higher-energy form, and **adenosine diphosphate (ADP)**, the lower-energy form. Enzymes are the means by which this energy is harnessed. Usually this is done by coupling some reaction P→Q that *requires* energy with a reaction like ATP→ADP, which releases energy. If you visualize the potential energy in a molecule as vertical position, you might think of this sort of enzyme as a sort of see-saw, in which one molecule's energy is increased, and another's is decreased, as in the figure below. (Dotted lines around a shape indicate a high-energy form of a molecule.)

Figure 16. A coupled reaction.

Cellular operations that require or produce energy will often use an **enzymatic pathway**—a sequence of enzyme-catalyzed reactions, in which the output of one step becomes the input of the next. One well-known example of such a pathway is the TCA cycle, which is part of the machinery by which oxygen and sugar is converted into energy and carbon dioxide. A small part of this pathway is shown below in Figure 17. (Notice that this particular pathway produces energy, rather than consuming energy).

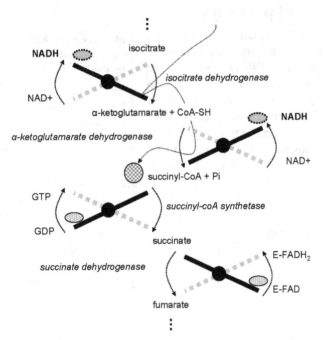

Part of the TCA cycle (also called the citric acid cycle or the Krebs cycle)
in action. A high-energy molecule of isocitrate has been converted to a
lower-energy molecule called α-ketoglutamarate and then to a still lower-
energy molecule, succinyl-CoA (as shown by the path taken by the
hashed circle). In the process two low-energy NAD+ molecules have been
converted to high-energy NADH molecules. Each "see-saw" is an
enzyme (named in italics) that couples the two reactions. The next steps
in the cycle will convert the succinyl-CoA to succinate and then fumarate,
producing two more high-energy molecules, GTP and E-FADH$_2$

Figure 17. Part of an energy-producing pathway.

Since each intermediate chemical in the pathway (e.g., fumarate, succinate, etc.) is different, each enzyme is also different: thus a pathway
that either consumes or produces large amounts of energy will often
involve many different enzymes, again contributing to complexity.

Amplification and pathways

Sometimes a pathway will act to amplify a weak initial signal. A good example of this is the pathway associated with **rhodopsin**. Rhodopsin is a G-linked protein receptor that detects light. Each rhodopsin protein cradles a "chromophore" molecule called **11-cis-retinal**. When a photon is absorbed by the 11-cis-retinal molecule, it changes shape, which causes rhodopsin to change shape and become "active." "Active" rhodopsin can then "activate" a second protein called **transducin**. Transducin, in turn, "activates" a third protein called **cGMP phosphodiesterase (PDE)**, an enzyme that hydrolyses a somewhat ATP-like molecule called **cyclic guanine monophosphate (cGMP)**. In the **rod** and **cone** cells in the retina—the

> The "fuel" used in a cell is chemically related to the bases of DNA and RNA. There are four **nucleobases** (aka **bases**) that form DNA: **adenosine, thymine, cytosine, and guanine**, abbreviated A, T, C, and G. (In RNA **uracil** replaces **thymine**.) A **nucleoside** is a base attached to a sugar: either **ribose** (for RNA) or **deoxyribose** (for DNA). A **nucleotide** is a nucleoside attached to a phosphate group: either mono-, di-, or triphosphate. These are abbreviated with 3- and 4-letter codes: e.g., ATP is adenosine triphosphate, and cAMP is cyclic adenosine monophosphate.

cells which sense light—cGMP acts somewhat like a chemical doorstop, propping open certain ion channels. When the concentration of cGMP is reduced, these ion channels close, changing the electrical charge of the cell and finally leading to a voltage signal. The process is thus something like this, where **R** is rhodopsin, **T** is transducin, and a* denotes the active form:

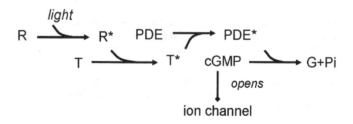

Figure 18. How light is detected by rhodopsin.

The interesting thing here, however, is that an active rhodopsin is unchanged after it activates a transducin, so it can go on and activate another transducin after the reaction completes. In fact, a single R* can activate thousands of transducin molecules per second, and likewise each PDE* can hydrolyze thousands of cGMPs per second. (A transducin can only activate one PDE, however.) This means that a single photon hitting the chromophore molecule can alter hundreds of thousands of cGMP molecules.

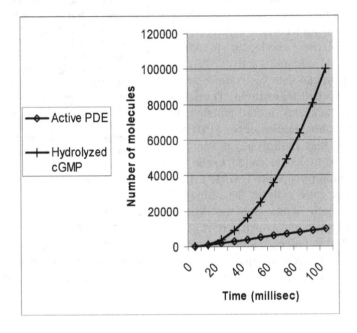

Number of molecules affected over time, assuming that each R* activates 100 transducin per second and each PDE hydrolyses 100 cGMP per second. (The actual numbers are larger).

The number of hydrolyzed cGMP molecules grows rapidly—at a quadratic rate—because it is product of two stages of "linear amplification". More stages of amplification would produce even steeper response curves.

Figure 19. Amplification rates of two biological processes.

In Figure 19, the pathway contains two "amplification" steps: both R*
and PDE* affect more than one molecule each. Notice that the number
of active transducin and PDE molecules grows linearly over time;
however, since each PDE* hydrolyzes a linear number of cGMP molecules per unit time, the number of cGMP hydrolyzed grows quadrati-
cally over time.

Modularity and locality in biology

Our understanding of macroscopic physical systems is guided by some
simple principles—principles so universally applicable that we seldom
think about them. One is the principle that *most effects are local*. This
means that a good start to understanding how something works is to
take it apart and see what touches what. Once we see that the ankle
bone connects to the shin bone, we understand that those two components are likely to interact somehow.

This sort of common-sense approach to understanding systems fails for
computer programs, where anything can affect anything. As a consequence, computer scientists are forced to construct elaborate schemes
to limit the interactions of software components—in Java, for instance,
private variables and methods, packages, and interfaces are all mechanisms for giving software constructs their own flavor of "locality."
Programs that do not observe these principles are notoriously difficult
to maintain, debug, and understand.

Like unconstrained software, the machinery of the cell also lacks
"locality." A bacterium, for instance, is a complex machine, with thousands of *types* of parts (the types of gene products) and millions of
instances of these parts. Although some of these parts form large
structures (like the flagella), many of them are essentially just sus-
pended in the fluid inside the cell. Components of the cellular machinery
find each other, interact, and then separate, often without preference
for a particular location.

This sort of non-local interaction is possible only for very small
objects, and at very small scales. In a bacterium, proteins move about
by **diffusion**, or random movement. In general, molecules around
room temperature move very fast: for instance, a molecule of air
moves at around 1000 miles per hour. However, molecules move

randomly, not systematically, which limits the ground that they cover. It is fairly easy to show that for objects moving by a random walk—specifically, objects that move a fixed distance in a random direction at each time step—the time it takes to cover a distance x with high probability varies as Vx^2, where V depends on distance traveled per unit time. This is very different from the macroscopic world, where the time to cover distance x is usually linear in x.

The result of this is that diffusion is a very quick way of moving around for very short distances—say, the width of a bacterium—and a very slow way of moving around over larger distances—say, from the bar to the buffet table. This may be why very little internal structure is necessary for bacteria, or for the bacteria-sized organelles in eukaryotic cells: there is simply no need for it, since everything is already close enough to interact quickly with everything else.

Over objects as large as a typical eukaryotic cell, however, simple diffusion is not necessarily the most efficient way for molecules to find each other and

> An **organelle** is a discrete component of a cell. Some but not all organelles are membrane-enclosed areas.

interact. For instance, the enzymes used by cells to digest sugar are all localized to the inner membrane of the mitochondria—they still move by diffusion, but in a limited, two-dimensional area.[1] The various membranes and organelles in eukaryotic cells, therefore, do not only *limit* the way that proteins interact, by isolating some proteins from others—they also may improve the speed at which interactions within that enclosure take place, by limiting diffusion to a small area.

Besides diffusion, eukaryotes have a number of other mechanisms for **transport**: for instance, **vesicles** are small organelles that move

---

[1] Very approximately, cell membranes are about the viscosity of butter, while the cytoplasm of a cell is about as viscous as water, so molecules move about 100 times as slowly when they are stuck in a membrane. However, diffusion in two dimensions is asymptotically more efficient than in three dimensions, so it is faster to diffuse inside a membrane if the distance is large enough. Analysis of simple model systems suggests that the "cross-over point" at which membrane-bound diffusion is faster than simple diffusion is somewhere between the size of a bacterium and a mammalian cell.

materials from organelle to organelle; and within the cytoplasm, some proteins are hauled from place to place along microtubules, which are long fibers that run radially from the center of the cell to the periphery. Transport in eukaryotic cells leads to locality, and hence to some degree of modularity, which can be used to help understand cellular processes. More generally, the **subcellular location** at which proteins are found is often an important indicator of function.

It should also be emphasized that, while membranes provide some notion of locality inside a cell, membranes allow small molecules to diffuse through them, and biological membranes also have numerous mechanisms to allow (or actively encourage) certain larger molecules to pass through. Furthermore, because of properties similar to the random-walk property of diffusion, molecules that come close to an organelle tend to remain close to it for a while, and brush against it many times—Figure 20 gives some intuitions as to why this is true.

The result of this is that if receptors for a protein p cover even a small fraction of the surface of an organelle, the organelle will be surprisingly efficient at recognizing p. As an example, if only 0.02% of a typical eukaryotic cell's surface has a receptor for p, the cell will be about half as efficient as if the *entire* surface were coated with receptors for p. Cell-sized objects thus have a "high bandwidth"—they can recognize or absorb hundreds of different chemical signals, even if they are bounded by membranes.

To summarize, understanding even the "simplest" living organisms is far from simple. Analysis of how the different components of complex biological systems relate to one another is usually called **systems biology**.

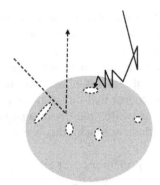

It can be shown that if a particle is released at distance δ from the surface of a sphere of radius R, it will touch the sphere before diffusing away with probability $p = R/(R+\delta)$. (See the book by Berg, 1983, cited in the last section, equations 3.1-3.5.) If the particle hits the sphere, bounces off, and returns to distance δ again, it has *another* chance to hit the sphere, again with probability p, so the expected number of times n it hits the sphere before diffusing away is

$$E[n] = \sum_{n=0}^{0} n \cdot \Pr(\text{exactly } n \text{ hits})$$

$$= \sum_{n=0}^{0} n \cdot p^n (1-p) = \frac{p}{1-p} = \frac{R}{\delta}$$

This means that a protein nearing a relatively large membrane-enclosed object (like a cell or organelle) is more likely to follow a path like the solid line than the dashed line—it will typically hit the cell many times before diffusing away, giving it many chances to "find" a receptor.

Figure 20. Behavior of particles moving by diffusion.

Looking at Very Small Things

Limitations of optical microscopes

The best way to understand and model complex systems is to obtain detailed information about their behavior. Biologists have developed many ways to obtain information about the workings of a cell. Some of these methods are clever and intricate, and many methods collect indirect evidence of behavior. I will start by discussing the most natural of these methods—the microscope—because, as Yogi Berra is reputed to have said, "you can observe a lot by just watching." For many purposes, the best way to study a cell is to look at it through a microscope.

Light microscopes have many advantages for biology. Relative to other sorts of radiation, light causes little harm to a cell—even highly focused laser light. Another advantage is that cells, which are largely water, are also largely transparent to light, which means that it is possible to look *inside* a living cell and watch it function. (The transparency of cells may come as a slight surprise to those that think of themselves as largely opaque. In fact, it is difficult to see through people only because they are many, many cells thick, and each layer of cell scatters a small amount of light.) Because of their transparency, cells are usually dyed in some way in order to be viewed in a microscope; this is more of an advantage than a chore, however, since there are many dyes that selectively color some parts of a cell but not others, thus emphasizing its structure.

One disadvantage of light is that objects that are too small simply cannot be resolved clearly with a light microscope. This limit is imposed by the wavelength of light. The wave nature of light implies that light waves interfere with each other, which distorts images: for instance, a point source of light will appear as a circle surrounded by a series of concentric circles. For some simple objects, one can precisely analyze the result of interference, and make precise claims about what can and what cannot be seen. Figure 21 summarizes one such result, which shows that wavelength λ of light

> The quantity $n \sin \theta$, where n is the **refractive index** of the medium being used, is called the **numerical aperture** of a microscope. Making the aperture wider improves resolution at the cost of depth of field.

and the **aperture** of a microscope—the width of the entry pupil—limits the amount of detail that can be distinguished for one class of simple objects.

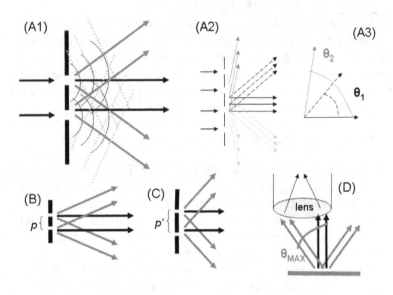

Abbe model of resolution: (A) Light passing through two pinholes propagates outward beyond the pinholes much as waves in water would (arcs in A1). Constructive interference between these waves (suggested by dotted lines) causes light to emerge only at certain angles (grey rays) called **diffraction orders**. A "perfect storm" for constructive interference of light with wavelength λ occurs when many pinholes are placed at a uniform distance p (A2); then the diffraction orders (A3) are at angles θ_1, θ_2, θ_3 etc, such that

$$p \sin \theta_N = N\lambda$$

Different spacings p,p' between the pinhole will lead to different diffraction angles (B), (C). To get enough information to determine the separation between pinholes, a microscope needs to capture rays from at least two diffraction orders. The **aperature** (width) of the microscope limits the angle between these to some θ_{MAX} and solving the equation above implies

$$p > \lambda/\sin \theta_{MAX}$$

Unless this holds, the two pinholes cannot be resolved.

Figure 21. The Abbe model of resolution.

(The figure ignores the issue of **refractive index**, which is the ratio of speed of light in the medium containing the specimen to the speed of light in air. Also, the limit outlined in the figure can be improved by a factor of 2 by considering light that enters the specimen at an angle.) Visible light has a wavelength of around 0.5 micrometers (μm), and objects smaller than 2 μm cannot be resolved even with the best light microscopes. This is adequate to resolve individual cells, and even the specialized organelles inside a cell, but not to visualize individual protein complexes or proteins.

A second disadvantage is that since cells are largely transparent, the signal obtained is fairly weak: put another way, the amount of light reflected by an object (or transmitted through an object) is not that much larger than the amount of light that is randomly scattered.

Special types of microscopes

One way to strengthen the signal is to use a technique called **differential interference contrast (DIC)**. Although only a small amount of light is reflected by an unstained cell, the refractive index of the cell is usually different from the surrounding medium: that is, light moves more slowly as it passes through a cell. This slight difference can be detected by comparing the phase of a light-wave that has passed through a cell with the phase of a light-wave that has not. A DIC microscope works according to this principle. (See Figure 22 below).

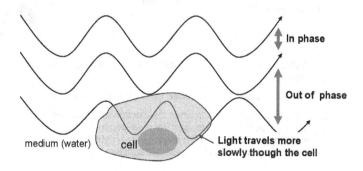

Figure 22. How a DIC microscope works.

Another way to obtain better images is with a **fluorescent** dye. **Fluorescent molecules** are molecules that *absorb* light of one frequency *f*, and very shortly after, *emit* light of another frequency *g*. This happens because when the atom absorbs light, an electron orbiting the nucleus of some atom in the molecule is pushed into a higher orbital—an orbit which is highly unstable. This unstable orbit typically lasts for a nanosecond or so, and when the electron returns to the lower, stable orbital, a photon is emitted. Importantly, the wavelength of the emitted photon is different from the wavelength associated with the absorbed photon, so that it is possible to filter out the reflections of the light which was intended to stimulate fluorescence, and detect very low levels of fluorescent light. In fact, it is possible to detect a very small number of fluorescent molecules (although one cannot form clear images of them).

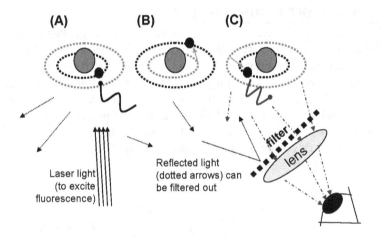

A. A photon is **absorbed**, pushing an electron to a higher-energy orbit.

B. The atom remains in an **excited state** for short time period.

C. The atom **emits** a photon when the electron returns to the low-energy orbit (dashed arrows). The wavelength of the emitted light is different from the wavelength of the laser light, so the emitted light can be separated from reflected light by a filter.

Figure 23. How a fluorescence microscope works.

Remarkably, it is now possible to create fluorescent dyes that are extremely specific—dyes that will bind themselves to only a few particular proteins in a cell—and use these dyes to visualize the behavior of specific proteins inside a cell. We will discuss two ways of doing this below, in the sections on antibodies and the section on fusion proteins.

Figure 24 shows some sample images from a fluorescent microscope. In this experiment, researchers were studying the behavior of a certain type of receptor protein called the HT-52A receptor, which is sensitive to a number of familiar substances including LSD, psilocin, and mescaline.

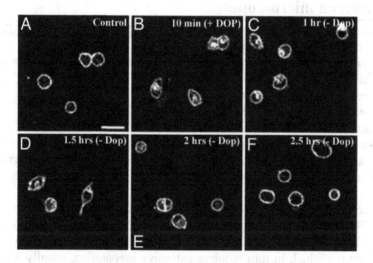

Fluorescent microscope images. These cells are cultured human cells, in which one of the G-couple protein receptors for serotonin has been made fluorescent. Panel (A) shows control cells, in which the fluorescence is all at the surface of the cell. Panel (B) shows cells that have been incubated with dopamine, a neurotransmitter, for 10 minutes. After exposure to dopamine, some of the receptors have moved to the interior of the cell—which suggests that the cell will be harder to stimulate with serotonin. Panels C-F show cells at various times after the dopamine has been removed: 1 hour, 1.5 hours, 2 hours, and 2.5 hours. After 2.5 hours, most of the receptors have once more moved to the surface of the cells.

(Reproduced from "Activation, internalization, and recycling of the serotonin 2A receptor by dopamine", by Samarjit Bhattacharyya, Ishier Raote, Aditi Bhattacharya, Ricardo Miledi, and Mitradas M. Panicker, *Proceedings of the National Academy of Science*, 2006, volume 103, pp. 15248-15253. Copyright 2006, National Academy of Sciences, U.S.A.)

Figure 24. Fluorescent microscope images.

One important type of microscope for use with fluorescent dyes is the **confocal microscope**, in which aggressive filtering is used, so that not only is reflected light filtered out, but also only light emitted by a very small part of the image is detected. A confocal microscope thus needs to scan progressively through a specimen. Good confocal microscopes can produce 3D images that include information from many different dyes. The confocal microscope was patented by Marvin Minsky (one of the founders of artificial intelligence) in 1957, but only became practical years later, with the development of lasers.

Electron microscopes

Electron microscopes use higher-frequency wavelengths, which gives them improved resolution, relative to optical microscopes. Electron microscopes can, in principle, resolve objects 10,000 times smaller than optical microscopes—in practice, however, current electron microscopes improve resolution by "only" a factor of 100. This makes it possible to see very small objects indeed. Figure 25 shows electron microscope images of a number of very small objects.

> **Mitochondria** are organelles that produce energy from glucose and oxygen. **Actin** is a protein that forms **microfilaments**, long filaments which help give a cell its shape.

Electron microscopes have some disadvantages, however. Electrons, unlike light, do not penetrate very far into a cell. Hence, if one wishes to visualize objects deep inside a cell, it is necessary to cut it into thin sections—which in turn requires extensive preparation, usually including dehydrating the cell and then allowing resin to permeate into it, or freezing the cell very rapidly. Electron microscopy also requires placing the cell in a vacuum, and staining the structures one wants to visualize with some sort of heavy element—e.g., gold. Both of these procedures (to put it moderately) tend to cause damage, so preserving a specimen in something like its native state is often a major challenge for electron microscopy. Work on using electron microscopes in close-to-normal conditions is an active area of research, however.

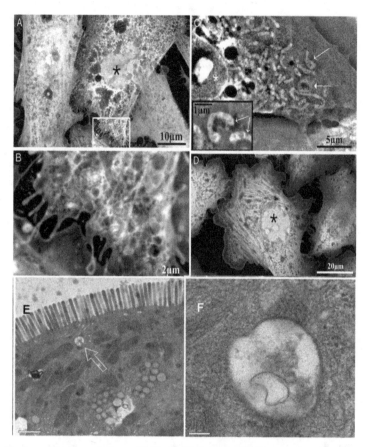

Electron microscope images. (A) Human HeLa cells. (B) The inset in (A), further magnified. (C) Hamster CHO cells, with some mitochondria shown in the inset. (D) Actin filaments. (E) Part of the intestinal cell of a 4-day old rat. (F) The vesicle indicated with an arrow in (E). Scale bars are 1 micrometer in E, 100 nm in F.

(Panels A-D are reproduced from Thiberge et al. "Scanning electron microscopy of cells and tissues under fully hydrated conditions", 2004, *Proceedings of the National Academy of Sciences*, volume 101,pp. 3346-3351. Panels E-F are from Lobastov et al. "Four-dimensional ultrafast electron microscopy", *Proceedings of the National Academy of Sciences*, 2005, volume 102, pp. 7069-7073. Copyright 2004,2005 National Academy of Sciences, USA.)

Figure 25. Electron microscope images.

Manipulation of the Very Small

Taking small things apart

A well-worn cliché is that cells are machines, the components of which are molecules. This leads to an important point: in general, molecules are too small to be seen or manipulated directly. How can one study a machine if you can't look at or manipulate its components? To use a computer science analogy to this problem, imagine trying to reverse engineer a PC from a hundred yards away, with your only tools for manipulation being a collection of bulldozers and excavators and such that you direct by remote control. What sort of things could you do, and what sort of things would you learn? Imagine for a moment a scientific field where a typical paper reads like this:

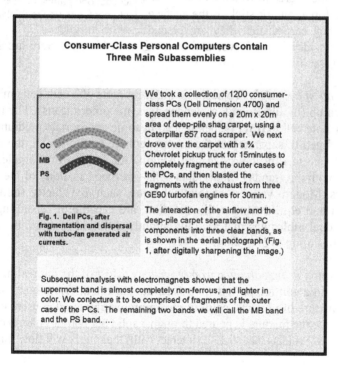

Figure 26. An article on reverse engineering PCs.

The study of the very small is analogous to this situation. In my face-tious example, we'd like to simply take the darn PC apart, but that's impossible to do with such crude tools; similarly, in biology, we don't have tools sufficiently delicate to disassemble cells directly, as we would disassemble a PC with hand tools. On the other hand, the crude PC "dis-assembly" of the example is far from useless: the authors *might* have successfully determined that PCs are, to a first approximation, made of an outer casing, a power supply, and a motherboard.

The general technique used in the exam-ple to separate PC components is to apply a force in one direction (air pressure from the turbofan) to a mixture. Elements of the mixture then get *separated* depend-ing on the degree to which they respond to the force, and/or stick to the surface (the shag carpet) that they are placed

> Splitting a mixture into components is called **fractionation** (if you're thinking about the input to the splitter) or **purification** (if you're using fractionation to collect one particular mixture element, and you're thinking about the output.).

on. This idea is used over and over again in biology. Here are some examples:

Separation by weight. To separate different parts of a cell, cells may be broken up (by ultrasound, a blender, or some other means). The resul-ting **whole cell extract** is then placed in some appropriate medium and centrifuged to separate out the components (e.g., the nuclei, the mito-chondria, etc.) by their size and weight. As in the PC example, one starts with thousands of individual cells—perhaps a colony of identical clones. Modern variants of this technique, such as **velocity sedimen-tation** and **equilibrium sedimentation,** are capable of separating out individual molecules that are only slightly different in mass, by using gravities of up to 500,000g.

Separation using column chromotography. Most of the interesting che-micals in a cell are proteins. To separate out the different components of a mixture of proteins, **column chromatography** is often used. In this technique the mixture is poured through a solid but porous column called a **matrix**. Proteins that stick to (interact with) the matrix will flow through the matrix slowly. By separating the fluids by the time that they take to emerge from the column, and choosing the appropriate matrix, proteins can be separated by size, electric charge, or **hydrophobicity** (i.e., affinity to water). The first types of column chromatography equipment took hours to perform this separation, but newer chromatography systems use tiny beads to form the matrix, and use high pressure to force a mixture through a column in minutes.

Separation by size or shape, using electrophoresis. Sometimes a matrix is placed on a flat surface, rather than a vertical column, and an electric force is used to move the components around, instead of a gravitational force. This technique is called **electrophoresis** and the matrix is called a **gel**. One very common gel method, especially for mixtures

> **Proteins** are linear sequences of molecules called **amino acids**. In a cell, this sequence will fold up into a complex shape, called the **tertiary structure** of the protein. The individual amino acids that make up proteins are sometimes called **residues**.

containing proteins, is **SDS-PAGE**, which is short for sodium dodecyl sulfate (SDS) polyacrylamide-gel electrophoresis. SDS is a detergent, which is mixed with the protein solution before adding it to the gel: it acts to unfold the proteins from their natural shapes into simple linear chains. The unfolded proteins migrate through the gel at a rate determined only by their sizes (not their shape after folding). A typical application of SDS-PAGE uses a single gel to compare several mixtures, each of which is placed in a different **lane** of the gel. An example of this is shown in Figure 27, below. It is possible to recover the proteins from a particular band of a gel (e.g., to run on a higher-resolution gel) by physically cutting out that band.

A variant of SDS-page is **2-D gel electrophoresis**. First, proteins are separated according to electric charge (using a special buffer in which pH varies from top-to-bottom) so that they spread out vertically in a narrow column. Then SDS is added to unfold the proteins; the ori-

> The technique for separating by charge used in 2D gels is called **isoelectric focusing**; it causes proteins to migrate to their **isoelectric point** (i.e., the point at which the protein has no net charge.)

ginal narrow columnar gel in placed on the left-hand-side of a wide SDS gel; and electrophoresis is used to spread the proteins out left-to-right according to size. Each protein will thus be mapped to a unique spot in two dimensions—unless of course there are two or more proteins with the same charge and size. A 2D gel can be used to separate 1000 different proteins, or in the hands of a master, even 10,000 proteins.

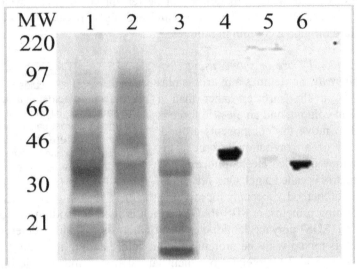

An example of an SDS-PAGE gel. Lanes 1-3 are a complex mixture of several substances, and lanes 4-6 show the corresponding mixture components after purification (via Western blotting, described below). The leftmost column is provided by the authors, and shows the molecular weights of substances that migrate to each level. Here the authors are demonstrating the effectiveness of the purification method used.

From *Mass spectrometric analysis of electrophoretically separated allergens and proteases in grass pollen diffusates,* by Mark J Raftery , Rohit G Saldanha , Carolyn L Geczy and Rakesh K Kumar, *Respiratory Research 2003, 4:10*

Figure 27. Using SDS-PAGE to separate components of a mixture.

Table 1. Different ways of sorting mixtures

Method	What is Typically Sorted	(Numeric) Property Sorted By
centrifugation	whole cell extract	weight
column chromatography	mixture of proteins	size, weight, charge, or hydrophobicity
gel electrophoresis	mixture of proteins or nucleic acids	size (folded) or electric charge
SDS-PAGE	mixture of denatured proteins	size (after denaturing)
2-D gel	mixture of proteins	size in one dimension, then electric charge in the second dimension

Separation by "affinity" to other substances. In all of the cases above, the "matrix" (i.e., the material through which objects move) interacts with the mixture components in a simple and fairly uniform way. For instance, in SDS-PAGE, the interaction between a protein and the gel can be described by one numerical parameter (protein size). Another application of separation-by-force is when the "matrix" has been designed to interact tightly with some elements, and not with others. An example is **affinity chromatography**, a variant of column chromatography in which one particular item in the mixture binds very tightly to the matrix, and other items do not interact at all. For instance, the beads of the matrix might be constructed so that they contain a particular strand of DNA, to which some unknown protein X in a mixture is suspected to bind. To isolate X, one simply pours in the mixture, and waits for all the other mixture elements to wash out. Finally one pours in some appropriate solvent that will break the bond to the mixture to obtain X.

To summarize, the types of fractionation that we've seen so far are the biologist's version of a common computer science operation: they **sort** mixture components according to a numeric function. Centrifugation sorts components by weight; gels sort components by size or charge. Affinity chromatography is a new type of operation, which extracts mixture components according to a "user-defined predicate"—it *selects* elements that pass a certain experiment-specific test.

> Biologists often use the term **selection** for a "user predicate that can be applied quickly, in parallel." For instance, one can **select** for antibiotic-resistant bacteria by treating a group of them with the antibiotic. A test that requires manual effort for each item is usually called a **screen**. To a first approximation, a **screen** is an $O(n)$ operation, and a **selection** is an $O(1)$ operation.

As another example of "user predicates" in biology, consider a situation in which we have a mixture M of many proteins, and a particular protein X that we know binds to some of the proteins in M. How could we determine which ones? Let us assume for a moment that we have some way of easily detecting X—for instance, we've done something clever so that X is radioactive, or perhaps it's been labeled to that it glows bright green. One possibility would be to construct a 2D gel M, and then use a sheet of slightly absorbent material to blot up the proteins in the gel, while preserving their relative positions. We now have a 2D arrangement of proteins which are *fixed* in position on the blotter. We then smear X evenly over the paper, and then carefully wash it off. Every location on the paper to which X sticks corresponds to a protein in M with which X interacts.

This technique is called a **Western blot**. Performing the analogous operations starting with a (one-dimensional) gel containing RNA molecules in order to determine which RNAs **hybridize** to some DNA molecule X is called a **Northern blot**. Performing a Northern blot with DNA instead of RNA is called a **Southern blot**. (Historically, Southern blots came first—they were invented by a biologist named Ed Southern in 1975.) The grandchild of the Northern blot is the infamous **gene chip** (and/or the closely related **microarray**), which I will talk about next.

It might be that two cells with the same DNA might build different sets of proteins—that is, they may have different **proteomes**. In single-celled organisms, a proteomic difference might be due to a response to different environments—for instance, different nutrients, or different temperatures. In a multi-celled organism, cells from different tissues express different sets of proteins. Studying such differences in genetic expression is a frequent goal in experimental biology.

Since proteins are always encoded by mRNA before they are built, one (indirect) way to detect differences in the proteomes of two cells is to compare their sets of mRNAs. In fact, this is actually a very convenient way to measure differences, because nearly all mRNAs have "**polyA tails**"—that is, the end of an mRNA is a long sequence of repetitions of the base **adenine**, which is abbreviated "A." The base complementary to adenine is thymine, abbreviated "T," and hence most mRNA will easily hybridize to a nucleic acid that consists of a long sequence of thymine residues. This means that one can easily use affinity chromatography to purify mRNA from a whole cell extract.

A **microarray** is an array of thousands of locations, each of which contains DNA for a different gene. Thanks to the magic of gene sequencing, VLSI-scale engineering, and robotics, a microarray that holds DNA for every one of the thousands of genes in yeast can be made and mass-produced fairly inexpensively, and is about the size of a microscope slide. A common use for microarrays is to take two mRNA samples from two cells (or more realistically, two cultures of similar cells) which one would like to compare, and using the "magic" of

> Both DNA and RNA can be either single-stranded, or double-stranded. In double stranded DNA/RNA, each strand is **complementary** to the other. In the right conditions and at the right temperature, two single strands that are complementary can spontaneously form a double-stranded molecule; this process is called **hybridization** or **base-pairing**. Hybrid strands can be DNA-DNA or DNA-RNA.

fluorescence tagging, dye the mRNAs in these samples red and green, respectively. Both samples are then spread across the microarray and allowed to **hybridize** to their corresponding genes—the positions of which are known on the microarray. Finally image processing is used to look at the color of each location.

Let's call the cells and associated samples *A* and *B*. For genes being expressed in both *A* and *B* to about the same extent, the corresponding microarray location will be yellow. Genes expressed in neither *A* nor *B* will have black locations. Genes expressed by A and not B will show as green, and genes expressed by B and not A will be red. The intensity of a color indicates the level of expression—that is, the number of mRNA molecules being transcribed.

A **gene chip** has a similar function, but different construction. The locations in gene chips contain shorter sequences of DNA—up to about 25 base pairs long—that are synthesized (or should we say fabricated?) right on the chip. Often the sequences are chosen so that there is a known, unambiguous mapping between these sequences and genes from some sequenced genome; in this case, gene chips can be used in the same manner as a microarray.

Gene chips and microarrays are new technology, but not a new technique: they are essentially just high-resolution versions of Northern and Southern blots. There is also a high-resolution analog of the Western blot, called a **proteome chip** in which many proteins are attached to a single chip—perhaps a transcription of every yeast gene—and which can be manufactured reliably[2]. The proteome chip is a more recent

Table 2. Methods for selecting components of mixture that satisfy some property.

Method	What is Selected	(Boolean) Property Selected For
affinity chromatography	mixture, e.g. of proteins	Does a mixture component bind to a user-selected substance?
Western blot or proteome chip	mixture of proteins	Does a protein bind to one of a set of user-selected proteins fixed on a substrate?
Northern blot, micro-array, or gene chip	mixture of RNAs	Does an RNA hybridize to one of a set of user-selected DNAs fixed on a substrate?
Southern blot, micro-array, or gene chip	mixture of DNAs	Does a DNA hybridize to one of a set of user-selected DNAs fixed on a subtrate?

---

[2] For more information on proteome chips, see *Global Analysis of Protein Activities Using Proteome Chips*, by Zhu *et al.*, Science 2001, Vol. 293, p.p 2101-2105.

arrival to the biologist's toolkit, but its impact may be ultimately comparable to that of microarrays and gene chips.

Parallelism, automation, and re-use in biology

At this point, let me take a break from the catalog of technical tricks, and make a few general observations about what we've seen so far.

We've seen that one occupation of biologists is developing ingenious ways to disassemble cellular-sized components. One great advantage of these techniques, which I have not emphasized so far, is that it is often easy to apply them in parallel, to many mixtures at once. As a computer scientist, I have been struck by the widespread use of this sort of "parallel processing" in biological experimentation.

In particular, all of the "blot-like" methods discussed above—Northern, Southern, and Western blots, microarrays, and gene and proteome chips—are naturally parallel. Consider a Western blot, which tests a protein X for interactions with the proteins on a blot: the experiment remains the same, whether the blot contains 100 proteins, 1000 proteins, or 10,000 proteins. If you like, the 2D gel functions as a 2D array of tiny little columns, just like those used in affinity chromatography—columns which can be easily used in parallel. More intriguingly, it is as easy to test a mixture of 1000 proteins $X_1, ..., X_{1000}$ against this "array of columns" as it is to test a single protein X! It is exactly this sort of parallelism that is exploited in a typical microarray experiment, in which *every* mRNA in mixture is tested for compatibility with *every* gene in a genome. (As an aside, this sort of parallel processing is also largely the reason that biologists are currently awash in experimental data—so much so that they are eager to get help interpreting it from long-haired former AI hackers like me.)

Gene chips and microarrays have another important property, which again makes biologists immensely more productive than they were a generation ago. In programming, the single biggest gain in productivity comes from software re-use, and, like a good software package, gene chips allow a sort of re-use. By this, I do not mean that an individual chip can be re-used—it can't. However, gene chips *can* be manufactured repeatedly at moderate cost, and hence the effort of designing and engineering them—the vast bulk of the total cost—can be amortized ("reused," if you like) over many related experiments. This is an important development, as doing a Western blot (or similar biological experimental procedures) requires technical expertise, practice, and some natural dexterity to accomplish successfully. This sort of human skill *cannot* be duplicated without expensive and painful processes (like postsecondary education). However, once one has overcome the large fixed cost of automating the procedure, the automated process *can* be duplicated—often at a surprisingly low cost.

Gene chips are just part of this trend—throughout biology, many experimental procedures are being automated, or partially automated. **Liquid-handling robots** can now carry out many routine procedures. In addition to the savings associated with automation, these robots are themselves parallel, in that they can dispense 8, 96, or even 384 fluids at once into in arrays of wells. This allows many operations to be performed at once.

Finally, although replicability of experiments is still important, many biological experiments not only produce replicable descriptions of the experimental procedure; increasingly, the biologists exchange *results* that others can build on directly, without having to first replicate the experiment that led to the result. These results are, in fact, re-usable resources, and many recent research projects are explicitly designed to construct such re-usable resources, rather than to address specific biological questions. In such projects, a lab will systematically perform all conceivable procedures of a particular type, and then make the results available as a service. An example of this sort of project might be the Yeast GFP Fusion Localization Database[3], which, among other things, provides researchers with a GFP-tagged variant of (almost) every protein expressed in yeast cells. Systematically repeating all possible variations of a process of this sort, and making the results available to other labs (in this case, as a series of GFP-tagged strains of yeast that can be purchased) means that subsequent researchers need never repeat this sort of procedure.

One could view this economically, as a move toward a "horizontal economy," in which each lab specializes so as to do a few things well. I

---

[3] http://yeastgfp.ucsf.edu/

prefer to view this from a programming prospective, and think of resources (like GFP-tagged yeast) as a sort of "subroutine package" for biological experiments. In programming, one might save time by using some other hacker's machine-learning software package; in biology, one might save time by using some other biologists' library of genetically engineered yeast.

Classifying small things by taking them apart

Let us now return to our discussion of experimental methods in biology. From the perspective of computer science, one way to describe the experiments discussed above is as various *implementations* of two basic operations:

1. Given an object X, take that object apart into components $W_1...W_n$, and then sort the components according to some numeric property $F(W_i)$.
2. Given an object X, take that object apart into components $W_1...W_n$, and then extract all components that satisfy some Boolean property $P(W_i)$.

Here X is usually a known object with unknown structure. One example of this generic task is centrifuging a whole cell extract to separate out the various components of the cell by weight. Another example is running a purified mixture of a cells mRNA over a gene chip in order to separate the individual mRNAs by their ability to hybridize to genes (and hence, to quantify the the amount of mRNAs in each separated population).

Another important class of tasks is the following:

3. Given an unknown object X and a set of known objects $Y_1,...,Y_n$, determine the Y_i that X is most similar to.

A good example of this sort of task is identifying a particular protein. Here X is the protein and the Y_i's are all possible proteins that could be expressed by the organisms from which X was isolated—e.g., if X was taken from a yeast cell, then the Y_i's might be the entire yeast proteome. Finding the most similar Y is a way of identifying X.

In information retrieval, a simple and commonly used way of measuring the similarity of two documents X and Y is to convert them to "bags of words," or counts of the number of times each word appears. More precisely, X will be represented as a function $h_X(w)$, where w is a word and $h_X(w)$ is the number of times w occurs in X. Well-known measures for the similarity of two functions can then be used to measure the similarity of h_X and h_Y—variations on an inner product being the most common. In short, a "bag of words" representation encodes a linear string X (the document) as a *histogram* of substrings of a particular sort (namely, words), and then uses histogram-based similarity metrics for comparison.

The same idea can be used to compare two proteins. First it is necessary to convert the protein, which is a single long sequence of amino acids, to a bag of "words." The usual way of doing this is to use some chemical that breaks up the amino acid sequence in a consistent, predictable way: for example, **cyanogen bromide** will break proteins after each **methionine** residue. Separating and sorting the fragments of the protein, using a gel or chromatography, will produce a specific pattern called a **peptide map**. Assuming that the "sorting" is done according to some function $f(z)$, where z is a fragment, one could formally represent the peptide map for protein P as a function $h_P(n)$, in which $h_P(n)$ is the number of fragments z in P such that $f(z)=n$. The peptide map is a "fingerprint" for the protein, and can be used to identify it from a list of candidates that have been previously "fingerprinted" by the same procedure.

The technique most widely used for the separation of protein fragments is **mass spectrometry**. The mass spectrometer sorts the fragments by their charge-to-weight ratio, and then counts the number of fragments of each size. The resulting histogram is called a **mass spectrum**. One advantage of mass spectrometry is that it can be used on extremely small amounts of a protein.

Another example of "classification by separation" is **DNA fingerprinting**. As with proteins, one begins by cutting the DNA into fragments using a chemical that cuts in a predictable way: for DNA, the chemicals that are used are called **restriction endonucleases** (which will be discussed more below). Since DNA sequences differ slightly from individual to individual, the "bag of fragments" representation of two DNA sequences are likely to be different. Such a difference is called a **restriction fragment length polymorphism** (or **RFLP**). Similarity between DNA sequences based on RFLPs is useful for several forensic purposes, such as testing parentage and identifying criminals. Fundamentally, RFLPs are no different from the other histogram-based similarity measures discussed above.

In DNA fingerprinting, usually only a subset of the fragments is considered. One particular approach is to look only at fragments that contain certain **minisatellites**. Minisatellites are portions of DNA that occur in repeating sections: more specifically, each minisatellite contains many repeating subparts, and the sections also appear multiple times in the genome. Often the number of repeats per section varies from person to person, as do the number and position of repeating sections. It is easy to find minisatellite-containing restriction fragments using hybridization, and the variations in repetition (and hence size) make the histograms of minisatellites-containing fragment lengths quite distinctive. These properties make them well-suited to DNA fingerprinting.

Reprogramming Cells

Our colleagues, the microorganisms

There is another whole family of approaches to studying very small objects: rather than attempting to study molecular-level processes with the (comparatively) huge and clumsy machinery that we humans can design, let us look for useful molecular-level tools we can find in nature. In particular, living cells are full of useful molecular-level machinery— what can we, as biologists, do with this existing machinery?

As it turns out, a huge amount of biological experimentation is based on either using "machines" that have been extracted from living cells, or by cleverly tricking living organisms to do some work for us. In this section we will discuss how some of this machinery is used.

Restriction enzymes and restriction-methylase systems

In nature viruses invade a cell by inserting a fragment of foreign DNA. One defense mechanism against viruses is a **restriction-methylase system** (R-M system). The first part of such a system is that DNA native to the cell is "marked" after it is produced. (Usually the marker is a methyl group attached to some of the nucleotides of the DNA, and thus the protein which adds this marker is called a **methylase**). The second part of the system is that unmarked DNA—i.e., DNA that has not been "modified"—is attacked by a complex molecule called a **restriction endonuclease** (**RE**), which cuts the DNA at certain specific sequences of nucleotides. For example, the RE named EcoRI "recognizes" the sequence "GAATTC," and cuts after the "G"; the RE named HaeIII cuts the sequence "GGCC" between the Gs and Cs; the RE

> A **nuclease** cuts nucleic acids, like DNA or RNA. Those that cut at the ends of a molecule are called **exonucleases** and those that cut in the middle are called **endonucleases. Restriction endonucleases** are named according to certain rules. The first three letters come from the organism from which the RE was obtained; an optional fourth letter identifies the "strain" of the organism; and the remainder is a Roman numeral. Thus, HindII (pronounced "hin dee two") is the first RE isolated from strain R_d of _Haemophilus influenzae._

named HindII cuts any sequence matching the regular expression "GT[TC][AG]AC" after the third nucleotide. (In each example above, the DNA being cut is double-stranded—e.g., EcoRI will cut DNA when it sees the sequence GAATTC on one strand, and the complementary sequence CTTAAG on the other).

It's important to realize just how sophisticated the machinery in a RE is. In spite of the fact that the *action* of a RE is easy to describe, the process involved in performing this action is quite complex. Consider, for example, the RE called BamHI. BamHI binds only to the DNA sequence "GGATCC." It is remarkable enough that the RE binds very specifically to this particular pattern, but on top of this, BamHI is a machine with moving parts—once it has acquired its target, it changes shape, as a prelude to cutting the DNA. The whole cleavage process requires no external energy (e.g., in the form of ATP) to accomplish.

One measure of complexity of an artifact to consider the effort that would be needed to re-design the artifact to work for a slightly different task. After years of study of natural REs, biochemists are only now beginning to understand how to modify REs so that they bind to different DNA sequences.

Fortunately, we don't need to completely understand the mechanism of an RE to use one. Just as one can use a complicated software module as a "black box" in programming, one can exploit an RE quite effectively, as long as we understand its "interface"—that is, what the RE does. We've already seen one common use of REs—they are used to create the RFLPs that are the basis of DNA fingerprinting. Another very important use is to construct new DNA molecules by "cutting and pasting" together strands of existing DNA.

Constructing recombinant DNA with REs and DNA ligase

It is clear how to cut DNA with an RE. But how does one "paste together" two slices of DNA that have been cut?

The answer to this depends in part on how the DNA has been cut. Recall that the RE EcoRI "recognizes" the sequence "GAATTC," and cuts after the "G." Recall also that complementary base pairs in DNA are A-T, for adenine and thymine, and G-C, for guanine and cytosine. The sequence "GAATTC" has an interesting property: if you reverse it, the resulting string "CTTAAG" is complementary.

Let's look at an example of a double-stranded DNA sequence containing the subsequence "GAATTC" and see how it gets cut by EcoRI:

Table 3. Small fragment of DNA before being cut by EcoRI

GATTACA	G	AATT	C	CATATTAC
CTAATGT	C	TTAA	G	GTATAATG

Here the grey area shows one fragment after the cut, and the white area shows the other. Notice that the resulting DNA fragments will be mostly double-stranded, but with single-stranded bits hanging off the end. The single stranded bits that stick out (AATT in the upper strand, and TTAA in the lower strand) are called **sticky ends**. Just as ordinary single-stranded DNA strands hybridize together, the sticky ends of DNA fragments cut with EcoRI will hybridize together. So, fragments of DNA cut by EcoRI can re-assemble themselves.

However, this re-assembly process is not perfect, as fragments can re-assemble in a different order. Consider a longer DNA molecule, with two EcoRI sites:

Table 4. A longer DNA fragment, showing how it is cut by an RE

GATTACA	G	AATT	C	ATTACCAT	G	AATT	C	CATATTAC
CTAATGT	C	TTAA	G	TAATGGTA	C	TTAA	G	GTATAATG

If there are two or more possible cutpoints, then first cutting a large population of DNA molecules with EcoRI and next allowing the resulting fragments to re-assemble will *not* be a null operation. Instead, since the fragments re-assemble randomly, the result will be *all possible* re-assemblies of the fragments. For the DNA above, this would include a molecule in which the middle section has been removed, producing the DNA molecule of Table 3.

(I have glossed over an important point here, which is that the re-assembled molecules are not structurally identical to normal double-stranded DNA. In normal DNA, the nucleotides in each strand are linked together by a strong type of molecular bond called a covalent bond, and the two strands are held together by weaker forces. When a RE cleaves DNA, the covalent bonds are broken, and they are *not* repaired if the DNA is re-assembled by hybridizing sticky ends together. However, the covalent bonds can be repaired by another naturally-occurring enzyme called **DNA ligase**).

> An **enzyme** is a protein that acts as a **catalyst**: that is, a protein that facilitates a chemical reaction, but is itself unchanged by that reaction. Most enzymes have verb-like names that end in "-ase." A **ligand** is a molecule that binds to specific place on another molecule, and joining two molecules together is also called **ligating** them together.

An important application of this sort of DNA cutting-and-pasting is to take two strands of DNA of the form xSy and wSz, where S is the sequence recognized by EcoRI (or any other RE that leaves "sticky ends") and w, x, y, and z are all different DNA sequences. Cleaving with the RE and then allowing the fragments to re-assemble will give the brand-new DNA sequences xSz and wSy. These sequences are **recombinant DNA**, and they have many uses.

Inserting foreign DNA into a cell

If DNA is the "programming language" for cellular behavior, then recombinant DNA molecules are new programs. An exciting question is: can one *execute* these new programs? Can one insert a synthesized DNA program in a living cell, and change its behavior? In short, can one "hack into" a cell?

Amazingly, this trick is not only possible, but a common procedure in experimental biology. In fact, there are several ways to insert "foreign" DNA into a cell. One approach is to take advantage of plasmids. Plasmids are common naturally in bacteria and less common in eukaryotic cells. Eukaryotic cells can be encouraged to accept plasmids in certain unnatural conditions, for instance, by mixing plasmids with cells in a salty solution, which makes cellular membranes somewhat leaky.

Thus, one way to "hack into" a cell is to use REs to create a new recombinant "program" and store it on a circular strand of DNA which contains an origin of replication—a plasmid—and then insert the plasmid into a cell. Since plasmids may or may not be accepted by a cell, a little extra work is necessary to determine which cells actually contain one. One approach is to start with a plasmid that when "unrolled" has the form *xyz*, where *xy* is a gene that contains the RE site *S* between *x* and *y*, and that also confers resistance to antibiotic A; and *z* is a gene that confers resistance to antibiotic B. Suppose you cleave this DNA and mix it with some DNA string *w* which was also cut from between two *S* sites: you will get plasmids of the form xw^Kyz, where *K* could

> When used like this, the plasmid is called an **insertion vector**. Plasmids were one of the first vectors that were used, but can only hold relatively short strands of DNA—say, 8000-20,000 base-pairs long. **Phages**—viruses that infect single-celled organisms—are also used as vectors. Phages can hold more DNA, and have evolved their own efficient mechanisms for inserting foreign DNA. Phages typically consist of nucleic acid (RNA or DNA) and a protein coat, and many of them self-assemble *in vitro*; thus it is only slightly more difficult to construct a phage containing recombinant DNA than a plasmid.

be zero (i.e., the original plasmid) or greater than zero (i.e., a plasmid containing your new gene). Notice that if $K>0$ then the gene *xy* is interrupted; in this case it will no longer function properly. Thus, randomly inserting this population of plasmids into cells will produce cells of three types: (1) cells that absorbed no plasmids, which will not be resistant to either A or B; (2) cells that absorbed a non-recombinant plasmid *xyz*, which will be resistant to both A and B; and (3) cells that absorbed a recombinant plasmid *xwyz*, *xwwyz*, etc., which will be resistant to B but not A. (Notice that all cells of type (3) will have similar behavior, since they contain equivalent sets of genes).

So, how can one select out cells that are resistant to B and sensitive to A? Resistance to B is easy to check: one can simply add B to the medium on which cell colonies are growing, and those that survive will be B-resistant.

> Cells that are not resistant to a drug are said to be **sensitive** to it. The technique described here for copying a group of cell colonies is called **replica plating**.

Checking for sensitivity to A is somewhat more complex. You first take a single petri dish D1 that contains many cell colonies, and copy the colonies, in parallel: one way to do this is to touch a blotter to the surface of the dish, and then touch the blotter to a new dish. After some cell growth, the result is a copy D2 of the colonies in D1, where all colonies have the same relative position in D1 and D2. You then treat D2 with A and see which colonies die off: these are sensitive to A. Finally you go back to the original disk D1 and pick out the colonies that were the sources of the A-sensitive colonies in D2.

It is common practice to use **selectable markers** (like resistance or sensitivity to antibiotics) to confirm that foreign DNA has been accepted in a cell. Most plasmids used as vectors now are engineered ones, containing many RE sites and many easily-selectable genes.

Genomic DNA libraries

Like most of the operations we've discussed, insertion of foreign DNA can be performed in a massively parallel way: for instance, one can use REs to create a large pool of plasmids representing different programs and then insert this pool, randomly, into a large number of cells. By controlling the relative amount of plasmid material and cells, one can ensure that with high probability, each cell will contain *at most* one plasmid, and by using selectable markers, one can select for cells that contain *at least* one plasmid.

One application of this method is to take an entire organism's genome— for instance, the complete DNA of a fruit fly—and, using REs, mechanical methods, or some mix of the two, break the DNA into pieces. This random collection of pieces can then be randomly inserted into cells, which are then grown in cultures such that each culture contains one foreign "piece." The collection of colonies is called a **genomic DNA library**. It is useful for many purposes—among them, finding the DNA code that gave rise to a particular mRNA molecule. (The DNA and mRNA can be quite different in eukaryotes, in which mRNA is spliced before it is translated). Since the colonies can replicate, of course, it is never necessary to "return" anything to this library—one can withdraw a copy of every piece, and the entire library will still be available for the next researcher.

Creating novel proteins: tagging and phage display

Another way that biologists "hack" organisms is to modify the genome so that new, unnatural proteins are produced.

> An organism used to produce proteins is called an **expression vector**.

For instance, hormones such as human insulin or human growth hormone can be synthesized by inserting the appropriate plasmid into some organism—usually a bacterium, a yeast cell, or a cultured mammalian cell. Since producing large quantities of a foreign protein is usually harmful, the typical approach is to insert a gene which can be easily **regulated**—that is, which has a promoter which is active only under certain easily-controlled conditions, such as when the temperature is high. Proteins are then synthesized by growing a large colony of the bacteria (or other expression vector) and then "turning on" protein synthesis by activating the promoter.

Another use of recombinant DNA methods is to *modify* some protein of interest. This might be done to test protein function in a very specific way—for instance, one might change the 432nd amino-acid residue of protein p to see if that residue is important to the function of p. It is also common to add an additional sequence of amino acids to the end of p in order to make p easier to recognize. For instance, there are certain short (8–12 amino acid) sequences that bind tightly to commercially available substances called antibodies (which are described further below). These antibodies can include (or can be combined with) fluorescent dyes, or tiny gold balls, enabling the protein to be seen in a fluorescence microscope or an electron microscope. Another use of antibodies is to isolate proteins, via affinity chromatography.

> The short sequence that "attracts" an antibody here is called an **epitope**, and the process described here is **epitope tagging**. Longer tagging sequences may contain multiple epitopes and/or special **cleavage sites** at which easily-available enzymes cut the sequence: these more complex epitope tags make it easier to purify the tagged protein. One commonly used tag is the **tandem affinity purification (TAP)** tag, which allows extremely accurate purification by a series of two steps of affinity chromatography using very gentle chemicals. Another common tag used for purification is **glutathionine S-transferase (GST)**.

Co-affinity purification is a variant of this in which one isolates a protein p of interest, as well as whatever proteins q,r,s might be bound to p—thus identifying p's potential binding partners.

Another type of marker that is often added to proteins is the sequence for **green fluorescent protein (GFP)**. As an aside, it should be noted that useful fluorescence is a rather rare property. The fluorescent dye molecule should fluorescence easily—otherwise, the large amount of light required to excite the molecule will damage the cell. The dye also should emit light at a very different wavelength from the excitation light—otherwise, it will be hard to filter out reflected emission-light "noise." Finally, the dye should not **photobleach** easily. **Photobleaching** occurs when a fluorescent molecule undergoes a structural change as a result of being excited—a change that prevents later fluorescence. This is a random event, which can happen with any fluorescent molecule: however with good dyes, photobleaching happens with low probability, so the dye will work for a long time.

Interestingly, although none of the 20 amino acids found in nature are fluorescent, nature has still managed to develop a very good fluorescent protein molecule. Green fluorescent protein is naturally found in a type of luminescent jellyfish. It is relatively small (238 amino acid residues) and very stable. The sta-

> A **fluorophore** is a molecule that fluoresces easily. Many fluorophores contain a long series of adjacent bonds that can collectively either contain n or $n+1$ electrons—the tyrosine ring in GFP being an example of such a structure.

bility is due to its shape: the GFP protein folds into a narrow tube, which protects a small group of amino acids that, after protein folding is complete, spontaneously form new covalent bonds and thus becomes a **fluorophore**.

Since the sequence for GFP is known, DNA hacking can be used to modify any protein p by appending the sequence for GFP—that is, by extending the **N-terminal** (or C-terminal) of p with the amino acid sequence for GFP. The modified protein (usually) has the same sort of behavior as the original protein, but can be easily seen using fluores-

> The combination of a target protein with a sequence from another protein is called a **fusion protein** or a **chimeric protein**.
>
> The two ends of a protein are called its **N-terminal** and **C-terminal** ends—see Figure 28 for an explanation.

cence microscopy. Variants of GFP that fluoresce in other colors are also available.

Yet another way in which fusion proteins are used is to modify a virus or phage so that the virus coat is combined with some protein p of interest—in other words, so the virus "displays" the protein p "on the outside," where it can easily bind with free proteins q. One powerful application of this **phage display** technique is to create a large library of phages, each of which "displays" a different protein p, grow large quantities of these phages and then mix them together to see which bind to the protein q of interest. It is usually relatively easy to isolate the phages that bind to q, and importantly, each phage contains its own DNA. This means that one can easily determine the identity of the binding protein p by sequencing the phage DNA.

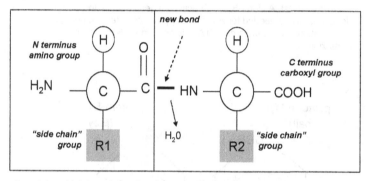

A protein, which is a chain of amino acids, has an **N-terminus** (where there is an unlinked nitrogen-containing **amino group**) and a **C-terminus** (with an unlinked carbon-containing **carboxyl group**).

Figure 28. Structure and nomenclature of protein molecules.

Yeast two-hybrid assays using fusion proteins

Recombinant **fusion proteins** can also be used to test to see if two proteins p and q bind to each other. The trick is to modify p and q by attaching tags that will make it obvious when and if p and q bind.

One natural setting where protein-protein binding is apparent is when that binding causes some gene to be transcribed. An example is shown on the top of Figure 29 below, where protein A

A **reporter gene** is one that will behave differently under different circumstances in an experiment, and which can be easily detected when it is expressed.

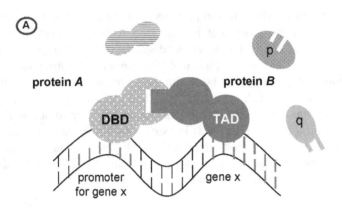

(A) In wild yeast, *A* binds *B*, which activates gene *x*. Only the DNA binding domain (DBD) is needed for *A* to find the promoter site, and only the transcription activation domain (TAD) is needed for *B* to activate transcription.

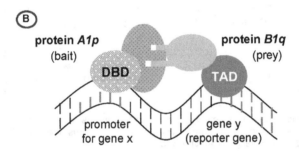

(B) In hybrid yeast, the DNA has a promoter for x near a reporter gene y. *A1p* can bind to the promoter site using the DBD of *A*, and *B1q* will activate transcription—of gene *y*—using the TAD of *x*. But *A1p* will only recruit *B1q* if proteins *p* and *q* bind. So, *y* is expressed iff *p* and *q* bind.

Figure 29. The yeast two-hybrid system.

recognizes the promoter for gene *x*, and then initiates the transcription of *x* by binding to a partner protein *B*. If the proteins *A* and *B* are somewhat modular, then one can exploit this natural setting to test for other protein-protein binding events. In particular, one can use recombinant DNA methods to test for binding if: (1) *A* consists of two independent parts, one which binds to the DNA, and one which binds to *B*, and if (2) *B* likewise consists of two parts, one which binds to *A*, and one which binds to the DNA.

The DNA-binding part of *A* is called the **DNA binding domain (DBD)**, and the DNA-binding part of *B* is called the **transcription activation domain (TAD)**. The natural scenario can be exploited by first modifying the DNA so that the promoter for *x* is paired with some gene *y* whose expression can be easily detected (for instance, *y* changes the cell's color when it is expressed). The DNA is also modified so that proteins *A* and *B* are replaced with two fusion proteins *A1p* and *B1q*. *A1p* combines *p* with the **DNA binding domain** of A—i.e., the part of *A* that actually binds to the promoter site. *B1q* combines *q* with the **transcription activation domain** of *B*, but not the part that binds to *A*. If *p* and *q* bind, then *y* will be transcribed and expressed, but otherwise it will not be. The process is shown on the bottom of the figure.

This process is usually carried out in yeast cells, and is called the **yeast two-hybrid system**. It lends itself to parallelism quite nicely: if there are many potential *p*'s and *q*'s, then it is possible to produce a population of yeast with alternative *p*'s, and a second population with alternative *q*'s, and then cross-breed the two populations. The children will each contain one of the *p* fusion genes, and one of the *q* fusion genes. The small numbers of child cells that contain a *p* and *q* which interact can be detected using the reporter gene.

Other Ways to Use Biology for Biological Experiments

Replicating DNA in a test tube

There are a number of reasons for wanting to insert foreign DNA into a cell, one of which is simply to **amplify** (increase) the quantity of foreign DNA by making use of a cell's natural abi-

> A molecule that contains a number of repeated units arranged linearly is a **polymer**. DNA, RNA, and proteins are all polymers.

lity to grow and multiply. However, there is a more direct way to amplify DNA, by using of some of the cell's DNA replication machinery *in vitro*. This technique is called **polymerase chain reaction (PCR)**. To explain how it works, I will first review, at a high level, how DNA is duplicated (**replicated**) in a cell. (The mechanisms for this differ somewhat in prokaryotes and eukaryotes—here I will focus on prokaryotic replication).

DNA replication consists of two main stages: **initiation**, and **polymerization**. In initiation, special proteins bind to the origin of replication in a double-stranded DNA, and separate the two strands. An enzyme called **RNA primerase** then builds two RNA molecules, each of which is complementary to a few bases of the separated strands. The RNA "**primers**" that are created by the primerase serve as a sort of scaffold that supports the next phase of initiation, in which a complex of proteins called a **replisome** is formed.

The replisome carries out the next stage of **polymerization**, in which the bulk of the DNA is duplicated. If DNA replication were an iterative program, the initiation phase would be the initiali-

> Figure 30 explains why the ends of DNA strands are called **3'** and **5'** (pronounced as "three prime" and "five prime").

zation, and polymerization would be the main loop. Polymerization of DNA is a complicated process, because single-stranded DNA is asymmetric: the two ends of each nucleotide are called the 5' and 3' ends, and nucleotides are always linked up 5' to 3'. In a double-stranded DNA, if the "top" strand is laid out with 5' on the left and 3' on the right (the usual orientation in textbooks), then the "bottom" strand would have the 5' on the *right* and the 3' on the *left*, like so:

Table 5. Dual-stranded DNA, with 5' and 3' ends labeled

5'   GATTACAGAATTCCATATTAC   3'
3'   CTAATGTCTTAAGGTATAATG   5'

In polymerization, the replisome moves in one direction (say, left-to-right)—which is quite a trick since the main work of duplicating the DNA is performed by three molecules of **DNA polymerase III**, an enzyme which repeatedly extends a partially duplicated DNA, but which only moves in the 5' to 3' direction.

To handle this difficulty, duplication along one of the two strands is performed with a series of jumps and re-starts: along the so-called "lagging strand" DNA is duplicated for runs of 1000 nucleotides or so in the *opposite* direction of replisome movement. Additional machinery is needed to patch up the discontinuities in the "lagging strand," to uncoil the DNA that is being replicated, and to "proofread" the generated DNA. This process is shown on the top of Figure 31.

To take a physical analogy, the replisome is like a car motor, and the initiation phase is like a starter motor—and the origin of replication is like the car keys. Starting the engine *in vitro* is difficult—as is assembling all the machinery needed to perform the replication process, including the jumps-and-restarts along the lagging strand. However, there is a way of "hotwiring" the replication process—i.e., initiating a (simplified) replication process *in vivo*.

The trick is to first *denature* the DNA, i.e., separate the double-stranded DNA into two complementary single-stranded molecules, and then

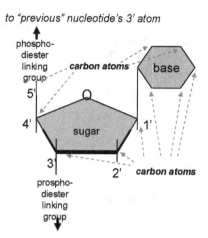

A **nucleoside** consists of a **nucleobase** (e.g., adenosine, thymine, cytosine, guanine) and a **sugar group**—ribose for RNA, and **deoxyribose** for DNA. Normally sugars are linear atoms, and the carbon atoms are numbered 1,2,3,4,5. In nucleic acids they fold into a ring, but the atoms are numbered in the same order; however, they are labeled 1',2',....,5' to distinguish them from the carbon atoms on the ring associated with the nucleobase (which are labeled 1,2,...,6).

A **nucleotide** is a nucleoside plus a phosphate group, which links it to the next nucleotide in the polymer. The phosphate groups link the 3' atom in one nucleotide to the 5' atom in the next. By convention, DNA strands are usually written with the "5' end" (the end with a "dangling" 5' carbon, not attached to any nucleotide) to the left.

Figure 30. Structure and nomenclature of DNA molecules.

combine these single-stranded molecules with short single-stranded "primer" DNA molecules that are complementary to part of it. After the primer hybridizes to a single-stranded DNA molecule, the result is a partially duplex DNA molecule—similar to the DNA molecules that are extended by DNA polymerase. The process is shown below on the bottom of Figure 31.

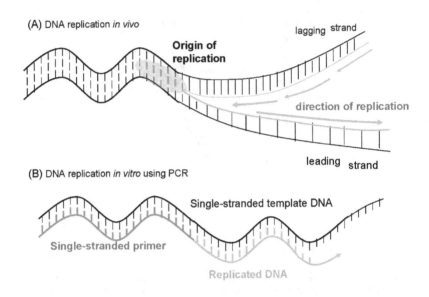

Figure 31. DNA duplication in nature and with PCR.

The complete PCR process is performed by first mixing the DNA that needs to be amplified with DNA polymerase, and a relatively large quantity of primer molecules. (The primers, or any other short DNA sequence, can be synthesized relatively easily by chemical means). One then (1) heats the mixture, which separates the two strands of the DNA; (2) cools it, allowing primers to hybridize to the separated single strands; and (3) waits for the DNA polymerase to replicate the remainder of each template-strand/primer pair. After this cycle, each single DNA strand has been turned into to a double-stranded version of itself— or to put it another way, the number of double-stranded DNA molecules has been doubled. One can then repeat steps 1,2,3 above again and again, doubling the amount of DNA in each cycle.

PCR is a very sensitive technique—it can be used to amplify even a single DNA molecule, which is very important for forensic purposes. The technique was made much more economical by the purification of DNA polymerase from "extremophile" bacteria that live in hot springs, as this flavor of polymerase is not damaged by the heat applied in each cycle.

Sequencing DNA by partial replication and sorting

One extremely important operation is determining the sequence of bases in a DNA molecule. Suppose that we can modify the DNA replication procedure described above so that it will occasionally stop at one of the four kinds of bases. (This can be done by adding to the PCR mixture above an appropriate quantity of a particular variant of the base called a **dideoxynucleotide**, which can be incorporated into a DNA being constructed, but halts replication after it has been incorporated). In fact, suppose that we can construct four "buggy" DNA-copying procedures, each of which occasionally stops a different one of the four nucleotides A,T,C,G (adenine, thymine, cytosine, and guanine). With these operations we can now sequence a strand of DNA, as follows.

First, provide the DNA to each of the four "buggy" copying procedures, and collect the results, in four separate test tubes. If you make enough copies, then you can be reasonably sure that you have all prefixes of the original DNA sequence, with prefixes that end in "A" in one test tube, prefixes that end in "T" in another, and so on. Performing this step on the string "GATTACA" might lead to the four populations of DNA shown in Figure 32.

Then, sort each population by size, using different **lanes** of the same gel. Again, the result is shown in Figure 32. The sequence can now be read off easily from the four lanes. Notice that you start reading the sequence at the *bottom*—normally substances are introduced at the top of a gel, and lighter molecules travel further.

Notice also that the *length* of DNA that can be sequenced is limited by the *precision* with which the molecules can be sorted by size. If only a 5% difference in size can be detected reliably, then only 20 bases can be sequenced at once; if a 0.001% difference in size can be detected, then 10,000 bases can be sequenced. In practice, about 1000 bases can be sequenced at once. This means that computational methods for pasting together many short overlapping sequences are needed to find the sequence of an entire organism.

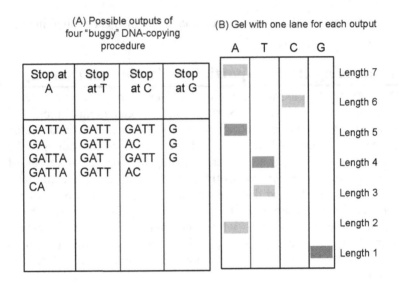

Sanger method for sequencing DNA. (A) The result of using variants of a DNA-copying procedure on many copies of a single strand of DNA. The variantsl randomly stop at prefixes that end in "A", "T", "C", and "G", respectively. The result of this is four populations of prefixes of the unknown DNA strand. (B) Using a gel to separate the four populations by weight. From the gel, the length-7 prefixes end in "A", the length-6 prefixes end in "C", and so on. Hence the final string can be reconstructed as "GATTACA".

Figure 32. Procedure for sequencing DNA.

This is one early method for sequencing DNA, and variants of it are still used today. One difference is that in modern sequencing methods, all four "buggy copies" are carried out in the same mixture, with fluorescent labels being added to the dideoxynucleotides to indicate the last base in a sequence. This makes it easier to automate the process of interpreting a sorted sequence of DNA fragments.

Other in vitro systems: translation and reverse transcription

A surprising number of complex biological activities can be performed *in vitro*. For example, the process of translating an mRNA into a protein can be carried out in a test tube: this is done by using (most of) a whole cell extract, which will contain a certain number of intact ribosomes. *In vitro* translation is a very useful way of finding what protein is generated by an mRNA: the trick is to add mRNA and radioactively-tagged amino acids to an *in vitro* translation system. The proteins built according to the template provided by the mRNA will be radioactive, which makes them relatively easy to find on a gel.

Another useful operation to perform on isolated mRNA is *reverse transcription*— conversion from an RNA to the corresponding DNA molecule. Reverse transcription is not part of the normal cycle of a cell, but certain viruses use reverse

> Reverse transcriptase is also used by **retrotransposons**, which are transposons that replicate themselves by using RNA as an intermediate.

transcription to infect a cell. Thus reverse transcription is performed by a naturally-occurring viral enzyme called *reverse transcriptase.*

DNA that has been formed from RNA by reverse transcription is called **cDNA**. One application of reverse transcriptase is to create another type of gene library,

> **cDNA** is short for **complementary DNA**. It is complementary to a mRNA molecule.

a **cDNA library,** by isolating (in parallel!) all the mRNA in a cell, and using reverse transcriptase to create the corresponding collection of cDNA. Unlike a genomic DNA library, in which every gene appears about once, a cDNA library will have *many* copies of genes that are expressed frequently, and *no* copies of genes that were not expressed in the organism from which the mRNA was drawn.

Exploiting the natural defenses of a cell: Antibodies

Mammals have **immune systems** that act as a defense against certain foreign substances. Our immune systems produce **antibodies**, complexes of proteins that bind very specifically to the foreign substance—that is, they bind to the foreign substance X, and to very few other things. Since antibodies bind so specifically, they are naturally useful in selecting

> An **antibody** is a protein complex that has been produced, by the immune system, so that it binds specifically to a particular **antigen** (foreign substance) X. The antibody might be called "antibody against X" or "anti-X."

out the substance X. A typical X might be some protein being studied.

To construct antibodies to a particular **antigen** X, one injects a small amount of X into some animal, usually a mouse or rabbit, and waits for the immune system to do its work. One then extracts some blood from the animal and extracts the **serum**—the fluid obtained by separating out the liquid part of the blood. This serum will be rich in antibodies, some (but not all) of which are antibodies for X. Isolation of the particular antibody for X requires an additional purification step.

If a larger quantity of an antibody is required, one possible procedure is to extract some of the cells in the mouse (or rabbit) that produce the antibodies. These cells, which are called **B-lymphocyte** cells, cannot be easily grown in culture, but they can be crossed (by certain unnatural means) with easily-cultured cancerous B-lymphocyte cells to create a particular kind of hybrid cell called a **hybridoma**. One can then screen for hybridoma cells that produce the desired antibody for X, and then culture these cells.

One important application of antibodies is to construct highly specific fluorescent dyes—dyes that affect only the protein X. This is typically done in a modular way with two types of antibodies: Ab1, an antibody against X, which is produced in (say) rabbits; and Ab2, a general-purpose fluorescently-tagged antibody against all rabbit antibodies. In the cell, Ab1 will bind to X, and Ab2 will bind to Ab1, thus tagging X.

> Using antibodies to fluorescently label antigens is called **immuno-fluorescence**. The analogous use of antibodies to make antigens visible to an electron microscope is called **immuno-EM**.

A similar process is sometimes used for electron microscopy, except that Ab2 will be tagged with small amount of heavy metal—for instance, a tiny sphere of gold—instead of a fluorescent tag.

Exploiting the natural defenses of a cell: RNA interference

To determine what a particular gene does, it is often useful to remove the gene completely from the genome. Often this can be done with recombinant DNA methods. However, sometimes there is a simpler way to achieve

> Removing a single gene from an organism is usually called **knocking out** the gene. "Silencing" a gene is often called **knocking down** the gene.

the same effect: sometimes it is possible to "convince" an organism that a particular gene should not be expressed, by using a mechanism called **post-transcriptional gene silencing (PTGS)**. PTGS can be used for many plants and animals—including fruit flies, nematodes, and several other widely-used **model organisms**. As the name suggests, genes are *transcribed* into mRNA, but the mRNA is "silenced" and never translated into proteins. The most common PTGS method is **RNA interference**, often abbreviated **RNAi**.

> Since many biologically important phenomena are **conserved** across many species, biologists often choose to work with **model organisms**—organisms that are particularly convenient to experiment on.

To use RNAi to silence a gene, one constructs a double-stranded RNA molecule that is complementary to the mRNA for the gene—or perhaps, one that is complementary to only part of the gene. Double-stranded RNA is not normally found in the cell—mRNA, for instance, is single-stranded—and it is attacked by an enzyme called **dicer** that chops it up into segments 21–23 bases long, called **siRNAs** (for "small interfering RNA"). Each of these siRNAs becomes part of a protein complex called an **RNA induced silencing complex** (RISC). The RISC, using the siRNA as a guide, finds and degrades any matching mRNA, thus preventing its translation to protein. The incorporation of the siRNA into the RISC complex makes this mechanism very specific: mRNAs that do not hybridize to the siRNA are left alone.

Gene silencing is not completely understood, but some of its uses in the cell are known—for instance, it is also used to regulate the expression of genes in some species. We also know that some viruses encode genetic material as double-stranded RNA, and RNAi thus acts as a defense against these viruses; in fact, this is probably how RNAi evolved. However, there is also a difficulty associated with the existence of RNA viruses: in mammals, long double-stranded RNA produce a strong antiviral response. To use RNAi in mammals, it is necessary to introduce the siRNAs directly.

Serial analysis of gene expression

Consider the way in which microarrays are usually used. The mRNA is extracted from a cell; cDNA is generated from the mRNA, using reverse transcription; the cDNA is amplified by PCR; and finally, a micro array is used to see which cDNA sequences are present in the amplified sample, and how common they are. An alternative to using a microarray is to simply sequence the cDNA, using the sequencing method discussed above. This direct approach could, in principle, be used to find gene expression levels in any organism, even one for which no microarrays are available; however, it is not used in practice, because sequencing that much cDNA is still quite expensive. However, there is a clever way in which cDNA levels can be measured by sequencing.

This method is called **SAGE**, for **serial analysis of gene expression**. The essence of the idea is to "summarize" the cDNA by snipping a random short sequence of nucleotides out of each cDNA strand—a sequence long enough to identify the gene, but not long enough to make sequencing impractical.

In the procedure adopted by SAGE, cDNA is generated from mRNA so that one end of it (the end complementary to the mRNA's polyA tail) is **biotinylated**—that is, marked with **biotin**, which binds readily to **avidin**. The cDNA is passed over a column of avidin-coated beads, which anchors one end of each cDNA on a bead. Then a restriction enzyme with a short (and therefore frequent) binding site is passed over the beads, cutting off some of the cDNA, and leaving the rest still bound to the bead.

The next step is to cut the beginning of the anchored segment (which is a random segment from *inside* the gene) off the bead, using a restriction enzyme such as FokI, which cuts several base-pairs "downstream" of a fairly infrequent binding site. Specifically, one first uses DNA ligase to "ligate" a special DNA sequence—let's call this sequence "marker A"—on to the free end of the anchored cDNA strands. Marker A contains a binding site for FokI, which then cuts the cDNA a short distance from marker A. The segments cut free by FokI all contain marker A, plus a small number of base pairs from the original cDNA, called a **tag**. We have now "summarized" each cDNA with a short subsequence, taken from somewhere in the middle of the gene.

Finally, in order to amplify the "summarized" cDNA, one repeats the procedure with a second marker sequence, marker B, and ligates together the results. Using appropriate primers, PCR will amplify only sequences between marker A and marker B. The amplified "summarized" cDNA tags are finally sequenced and matched against the known genome for the organism being studied; this shows which genes were originally present in the sample. The process is shown in Figure 33.

The SAGE technique illustrates an important point. Many of the procedures discussed above involve using a number of different techniques in sequence to achieve some effect. One of the subtleties of designing biological experiments is that designing an experiment—particularly one that involves using several techniques—requires a thorough understanding of the reliability and cost of different techniques. Naively, the "summarization" step used in SAGE is not necessary: in practice, however, it is essential.

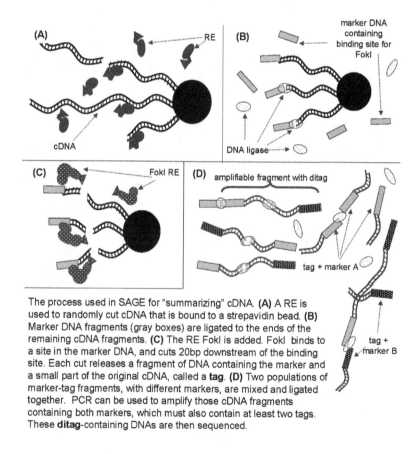

The process used in SAGE for "summarizing" cDNA. **(A)** A RE is used to randomly cut cDNA that is bound to a strepavidin bead. **(B)** Marker DNA fragments (gray boxes) are ligated to the ends of the remaining cDNA fragments. **(C)** The RE FokI is added. FokI binds to a site in the marker DNA, and cuts 20bp downstream of the binding site. Each cut releases a fragment of DNA containing the marker and a small part of the original cDNA, called a **tag**. **(D)** Two populations of marker-tag fragments, with different markers, are mixed and ligated together. PCR can be used to amplify those cDNA fragments containing both markers, which must also contain at least two tags. These **ditag**-containing DNAs are then sequenced.

Figure 33. Serial analysis of gene expression (SAGE).

Bioinformatics

As biologists become better and better at collecting data, the problem of managing, analyzing and interpreting the massive amount of data that has already been collected becomes more and more important. The growing field of **bioinformatics** is focused on this problem; more broadly, it is concerned with using computational methods to solve problems from biology. There are now a number of review articles, books and even complete educational curricula on bioinformatics, so this chapter is not in any way complete: however, it will hopefully give a useful overview of what the most active topics are.

One type of information that is available in large quantity is DNA sequences. The genomes of more than 180 organisms have been sequenced to date, and the nucleotide sequence for the human genome alone contains over 3 billion base pairs. One practically important problem is "googling" the database of genomic information—i.e., finding items of interest in a collection of sequenced DNA. A typical task might be finding possible **homologs** of some particular gene of interest. From a programmer's point of view, this amounts to the problem of finding, inside some very long string S, all substrings T that are "similar" to a "query string" Q: here S is the sequence database, Q is the DNA for the gene of interest, and the T's are the homologous "target" genes.

> Two genes from different organisms that are highly similar in sequence are **homologous. Homologs** from the same organism are called **paralogs**, and **homologs** from different organisms are **orthologs**.

The first step in solving this problem is to define what is meant by "similar." One simple definition for similarity is the **minimal edit distance** between Q and T relative to some set of **edit operations**. As an example, consider the operations *delete*, *insert* and *substitute*, which correspond to deleting a single

> **Levenshtein** distance assigns a unit cost to each edit operation. In aligning amino acid sequences one usually assigns different costs to different substitutions: this variant is called **Needleman-Wunch** distance.

letter, inserting a single letter, and changing one letter to another, respectively. For instance, the string "will cohen" can be changed to "walt chen" with two substitutions and one deletion.

There is a elegant method for computing the minimal edit distance between two strings Q and T in time $O(|Q|*|T|)$. The method takes advantage of the following recursive definition for the minimal edit distance between the first m letters of Q and the first n letters of T:

$$\text{distance}(Q,T,m,n) = \min \begin{cases} \text{distance}(Q,T,m-1,n)+1 & \text{// insert} \\ \text{distance}(Q,T,m,n-1)+1 & \text{// delete} \\ \text{distance}(Q,T,m-1,n-1)+1 & \text{// substitute} \\ \text{distance}(Q,T,m-1,n-1) & \text{if } Q_{m-1} = T_{n-1} \end{cases}$$

It's fairly easy to see why this definition works: for instance, the third line results from recursively finding the minimal edit distance between the first $m-1$ letters of Q and the first $n-1$ letters of T, and then substituting T_{n-1} for Q_{m-1} at an additional cost of one edit operation.

A naïve implementation would be slow, but one can compute the definition efficiently using dynamic programming. Alternatively one could "memo-ize" the function above—i.e., one could build a cache for each pair of arguments Q, T that saves the results for each m, n pair so that it need only be computed once.

This computation is shown in Figure 34: the figure shows the Levenshtein distance between the strings "will cohen" and "walt chen." Each entry in the matrix can be computed by looking only at entries above and to the left of it. The final distance between the two strings appears in the bottom-right corner of the matrix—in this example, the distance is 3.

There are many types of edit distances. One is **Smith-Waterman**, which is most easily described as a similarity measure, rather than a distance. It is defined by this recursive function:

	w	a	l	t		c	h	e	n
w	0	1	2	3	4	5	6	7	8
i	1	1	2	3	4	5	6	7	8
l	2	2	1	2	3	4	5	6	7
l	3	3	2	2	3	4	5	6	7
	4	4	3	3	2	3	4	5	6
c	5	5	4	4	3	2	3	4	5
o	6	6	5	5	4	3	3	4	5
h	7	7	6	6	5	4	3	4	5
e	8	8	7	7	6	5	4	3	4
n	9	9	8	8	7	6	5	4	3

An example of how to compute the Levenshtein distance between two strings. The *i,j*-th element of the matrix stores distance(Q, T, i, j), and the value of the lower right-hand corner entry (i.e., 3) is the distance between the two strings. The shaded entries are those that were used in the computation of the minimal cost (i.e., the cases of the *min* computation that were used to find the final score).

Figure 34. Computing a simple edit distance.

$$\text{score}(Q,T,m,n) = \max \begin{cases} 0 & //restart \\ \text{score}(Q,T,m-1,n)-1 & //insert \\ \text{score}(Q,T,m,n-1)-1 & //delete \\ \text{score}(Q,T,m-1,n-1)-1 & //substitute \\ \text{score}(Q,T,m-1,n-1)+2 & \text{if } Q_{m-1} = T_{n-1} \end{cases}$$

This scoring function gives a reward of 2 for "matching" at a single character position, a penalty of 1 for an insert, delete, or substitution, and unlike the Levenshtein distance above, allows the score to be "reset" to zero at any point. The final value used for score(Q,T) is the maximum value for score(*Q*,*T*,*m*,*n*) over all *m* and *n*. (The numbers used for rewards and penalties chosen here are picked for simplicity— other values, more appropriately reflecting the cost of changes, would be used in a real application). Figure 33 shows an example of this computation. Notice that the ability to "restart" at zero means that high scores can reflect a *partial* match between the two strings.

In the figure, I have shaded the "locally maximal" scores—scores with no higher-scoring neighbor—and the values that were used in the series of "max" computations leading to these locally maximal values. The shaded areas tend to be approximately diagonal, and if you look at the strings directly above or below them, you can identify the strings parti- cipating in the partial matches, and determine exactly where substitu- tions and deletions took place, according to the optimal edit sequence: for instance, you can determine that "will cohen" partially matches "walt chen" with a score of 12, and that the first "i" was in "will" was replaced by an "a" in "walt." This match is called an **alignment**.

	w	i	l	l		w	a	l	t		c	h	e	n		c	o	m	e
w	2	1	0	0	0	2	1	0	0	0	0	0	0	0	0	0	0	0	0
i	1	4	3	2	1	1	1	0	0	0	0	0	0	0	0	0	0	0	0
l	0	3	6	5	4	3	2	3	2	1	0	0	0	0	0	0	0	0	0
l	0	2	5	8	7	6	5	4	3	2	1	0	0	0	0	0	0	0	0
	0	1	4	7	10	9	8	7	6	5	4	3	2	1	2	1	0	0	0
c	0	0	3	6	9	9	8	7	6	5	7	6	5	4	3	4	3	2	1
o	0	0	2	5	8	8	8	7	6	5	6	6	5	4	3	3	6	5	4
h	0	0	1	4	7	7	7	7	6	5	5	8	7	6	5	4	5	5	4
e	0	0	0	3	6	6	6	6	6	5	4	7	10	9	8	7	6	5	7
n	0	0	0	2	5	5	5	5	5	5	4	6	9	12	11	10	9	8	7

Computing the Smith-Waterman similarity between two strings. The largest element of the matrix (i.e., 12) is the similarity. The long shaded area is associated with the score 12, and the substrings "will cohen" and "walt chen". The other shaded areas correspond to an exact match of the substring "will_" (with a score of 10) and an approximate match of "_cohe" to "_come" (with a score of 7).

Figure 35. The Smith-Waterman edit distance method.

In the example, the Smith-Waterman computation locates the target T="walt chen" as the best substring matching the query Q="will cohen" in the longer sequence S="will walt chen come." Many of the tools biologists use to find proteins are much more sophisticated, but based on the same underlying principle.

Similarities between genes can be exploited in other ways. For instance, human hemoglobin is more similar to mouse hemoglobin than sparrow hemoglobin, and more similar to sparrow hemoglobin than shark hemoglobin. Intuitively, this pattern of similarities makes the evolutionary tree (A) more likely than (B) in the figure below. There has been much work on the computational question of how to properly formalize this intuition, and how to efficiently search for the most likely evolutionary tree given a particular formalization.

> The study of evolutionary history is called **phylogeny**, and the trees shown in Figure 36 are often called **phylogenetic trees**.

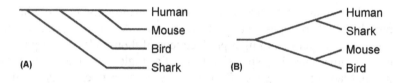

Figure 36. Two possible evolutionary trees.

In some cases, the rate at which proteins change over time can be inferred from comparing evolutionary trees to the fossil record. The inferred rates of evolution can then extrapolated to determine when other species diverged—even species not well-represented in the fossil record. Widely-conserved gene products like ribosomal RNA can thus be used as "**molecular clocks**" to estimate the rates of slower evolutionary processes; likewise, more quickly-evolving proteins are useful in estimating fast evolutionary processes.

"Data mining" sequence databases to find interesting regularities is also another important area of bioinformatics. Many large proteins are to some degree modular, and the modular subsequences are called **domains**. (An example is the DNA binding domain used in yeast two-hybrid assays). Protein domains

> A **domain** is a modular component of proteins: i.e., a subsequence that is approximately duplicated in many different proteins, and has approximately the same structure whenever it occurs. A **motif** is a small domain.

and **motifs** are one type of regularity that can be discovered by computational means. It is worth noting that when performing this sort of data-mining, a crucial computational decision is how to *represent* a discovery. For example, assume that different instances of a protein domain can all differ by a few amino acid positions, and that no single amino acid is always the same: how do you define such a domain computationally? One increasingly popular choice is to adopt a probabilistic framework, in which the "definition" of a domain is associated with some sort of probability model which describes how instances might vary.

Probabilistic and statistical methods are also widely used to help interpret the results of high-throughput experiments. As an example, a single microarray experiment might produce tens of thousands of data-points, each summarizing the expression level of a single gene in a single condition. Many of these genes will show different levels of expression under different conditions; however, it is quite difficult to determine which of the many changes in expression-levels result from chance fluctuations or experimental error, and which reflect some biologically interesting fact. Development of statistical techniques to analyze such high-throughput experiments is an active area of research, and the techniques proposed range from relatively simple analysis steps—such as testing to see if a particular pair of genes are likely to be co-regulated—to automatically constructing complete models of biological pathways.

Development of tools for helping biologists monitor, browse, and search the scientific literature is another active area of research. There are millions of scientific articles already in the literature, and the rate of publication has been steadily increasing in recent years. There are many active projects devoted simply to distilling this information into more readily accessible forms: e.g., databases describing known protein-protein interactions. Building such **curated** databases is expensive, as it requires human effort to read and understand biological publications.

Use of natural language processing methods and machine learning techniques to even partially automate the curation process could greatly reduce its cost.

Where to go from here?

This document is aimed at computer scientists who are trying to acquire a "reading knowledge" of biology. For those that want to learn more about core biology, the gentlest introduction I know of is "The Cartoon Guide to Genetics." The most comprehensive introduction is "Molecular Biology of the Cell, 4th Edition," by Alberts et al., which also has the virtue of being freely available on-line at the National Library of Medicine (NLM). If you're a non-biologist hoping to get along in biology, you could do worse than to read the former, and skim through the latter:

- **The Cartoon Guide to Genetics** (1991) by Larry Gonick and Mark Wheelis. Published by HarperCollins.

- **Molecular Biology of the Cell** (2002) by Bruce Alberts, Alexander Johnson, Julian Lewis, Martin Raff, Keith Roberts, and Peter Walter. Published by Garland Publishing, a member of the Taylor & Francis Group.

There is also a plethora of on-line information. Another gentle introduction to biology is "Molecular Biology for Computer Scientists," a chapter in a book entitled "Artificial Intelligence and Molecular Biology," edited by Lawrence Hunter, which is currently available on-line at http://www.aaai.org/Library/Classic/hunter.php. Several texts, including a complete copy of the Alberts *et al.*, textbook—all 1600 pages!—are also on-line at the National Library of Medicine, at the following URL: http://www.ncbi.nlm.nih.gov/entrez/query.fcgi?db=Books.

One visually appealing online resource is the collection of Flash animations on http://johnkyrk.com. There are also several hyperlinked textbooks, one of which is available from MIT at http://web.mit.edu/esgbio/www/. Dictionary.com is also a surprisingly good resource for finding technical definitions.

For persons interested in text-processing applied to scientific, biological text, some useful sites include these:

- BioNLP at http://www.ccs.neu.edu/home/futrelle/bionlp/

- BioLink at http://www.pdg.cnb.uam.es/BioLink

- BLIMP at http://blimp.cs.queensu.ca/.

There is also a good recent review article on NLP and biology, by Aaron Cohen (no relation) and Larry Hunter. Another recent review article, coincidentally by Jacques Cohen (again, no relation!) surveys **bioinformatics**, rather than biology.

- *Natural Language Processing and Systems Biology*, by K. Bretonnel Cohen and Lawrence Hunter. In Artificial Intelligence and Systems Biology, 2005, Springer Series on Computational Biology, Dubitzky W. and Azuaje F. (Eds.). This paper can also be found on-line at http://compbio.uchsc.edu/Hunter_lab/Cohen/Cohen.pdf.

- *Bioinformatics—An Introduction for Computer Scientists,* by Jacques Cohen, in ACM Computing Surveys, 2004, vol. 36, pp. 122–158.

In preparing this I used several additional textbooks and/or web sites as references:

- **Biochemistry** (2002), by Mary K. Campbell, and Shawn O. Farrell. Published by Thomson-Brooks/Cole. A good introductory textbook on biochemistry.

- **Biological Sequence Analysis: Probabilistic models of proteins and nucleic acids** (1998), by R. Durbin, S. Eddy, A. Krogh, and G. Mitchison. Published by Cambridge University Press. An excellent introduction to the many aspects of sequence modeling, including hidden Markov models, edit distances, multiple alignment, and phylogenetic trees, this text has uses beyond biology as well.

- *An Introduction to the Genetics and Molecular Biology of the Yeast Saccharomyces cerevisiae* (1998), by Fred Sherman. On the web at http://dbb.urmc.rochester.edu/labs/sherman_f/yeast/. This web site is a detailed description of yeast, a popular model organism for genetics. It is a modified (presumably updated) version of: F. Sherman, Yeast genetics, In The Encyclopedia of Molecular Biology and Molecular Medicine, pp. 302–325, Vol. 6. Edited by R. A. Meyers, VCH Pub., Weinheim, Germany, 1997.

- **Molecular Biology, Third Edition** (2005), by Robert F. Weaver. Published by McGraw-Hill. This book contains many in-depth discussions of the research, results, and reasoning processes behind our understanding of biology, illustrated by detailed analysis of specific research papers. It is a good resource for those wanting to obtain a "reading knowledge" of biology— that is, for those that want to be able to read and understand recent publications in biology.

- **Random Walks in Biology** (1983), by Howard Berg. Published by Princeton University Press. This is a short book with some very accessible discussions of diffusion in biological systems.

- **Transport Phenomena in Biological Systems** (2004), by George Truskey, Fan Yuan, and David Katz. Published by Pearson Prentice Hall. An in-depth treatment of transport and diffusion.

- **Biological Physics: Energy, Information, Life** (2003), by Philip Nelson. Published by W.H. Freeman. A beautiful and very readable treatment of the mathematics behind a number of biologically important processes, including diffusion, energy transfer, self-assembly, and "molecular machines."

Acknowledgements

I would like to thank Susan Cohen, for indexing the book, and encouraging me to write it; Dan Kundin, for proofreading a late version of the book; Eric Xing, for comments on an earlier version; and the National Institutes of Health, for supporting this work under NIH Grant DA017357–01.

Index